David Murry

An Archaeological Survey of the United Kingdom

David Murry

An Archaeological Survey of the United Kingdom

ISBN/EAN: 9783337243418

Printed in Europe, USA, Canada, Australia, Japan

Cover: Foto ©berggeist007 / pixelio.de

More available books at **www.hansebooks.com**

An Archaeological Survey of the United Kingdom: The Preservation and Protection of Our Ancient Monuments.

An

Archaeological Survey

of the

United Kingdom

By

David Murray, LL.D., F.S.A.

"Tumulus et lapis sint in testimonium."

Glasgow

James MacLehose and Sons

Publishers to the University

1896

"What is aimed at is an accurate description—illustrated by plans, measurements, drawings or photographs, and by copies of inscriptions—of such remains as most deserve notice, with the history of them so far as it may be traceable, and a record of the traditions that are retained regarding them." Minute by Lord Canning, Governor-General of India, in Council, 22nd January, 1862. Cunningham, *Archaeological Survey of India*, vol. I., p. 111 (Simla, 1871, 8vo).

"Je m'appliquerai à faire dresser un inventaire complet, un catalogue descriptif et raisonné des monumens de tous les genres et de toutes les époques qui ont existé ou existent encore sur le sol de la France." Guizot (1831), *Rapports et pièces*, p. 23.

"There is at the present time a most pressing want for an archaeological survey of Great Britain, including within the scope of its operations the plotting down upon the Ordnance Map of every trace of man and his handiwork left by successive generations upon the face of the country." J. Romilly Allen (1889), *Journal Brit. Archaeol. Assoc.*, xlv., p. 299.

"It is, no doubt, in a great measure owing to an ignorance of the extensive and rich archaeological fields which are scattered over the surface of these islands, that so little interest in the protection of these records of early civilization has been manifested by the Legislature, and by the public generally." Earl of Carnarvon (1881), *P.S.A.*, 2nd S., viii., p. 515.

"L'archéologue, en effet, cherche non-seulement à connaître et à décrire les œuvres des siècles passés ; il doit encore perpétuer le souvenir de ces monuments que l'âge atteint avec plus ou moins de rapidité et qui disparaîtront tôt ou tard ; il faut donc que l'archéologue, en même temps qu'il les décrit, représente les sujets qu'il veut faire connaître." Trutat, *La Photographie appliquée à l'Archéologie*, p. 1 (Paris, 1879, 12mo).

"A grave mound, a lonely circle of stones, a stone implement, a metal ornament excavated from the covered chamber of death, afford a livelier image of antiquity than Saxo or Snorre, the Eddas, or the Germany of Tacitus." *Guide to Northern Archaeology*, p. 25 (London, 1848, 8vo).

PREFACE.

The following pages are a reprint, from the *Transactions*, of a presidential address delivered at the opening of the last session of the Archaeological Society of Glasgow. It is issued in the present form in the hope of directing attention to the necessity of having an archaeological survey of the United Kingdom carried out by Government on lines similar to those of the topographical and geological surveys, and of further legislation for the protection and preservation of our Ancient Monuments.

DAVID MURRAY.

169 West George Street,
    Glasgow, 23rd *July*, 1896.

CONTENTS.

	PAGE
ARCHAEOLOGY A SCIENCE,	11
AN ARCHAEOLOGICAL SURVEY,	19
NATURE OF THE SURVEY,	29
THE PROTECTION OF OUR MONUMENTS,	45
THE ANCIENT MONUMENTS PROTECTION ACT,	47
THE PROTECTION OF MONUMENTS IN OTHER COUNTRIES,	50
FINDS; TREASURE TROVE,	57
PRESERVATION OF ANCIENT MONUMENTS,	71
RESTORATION,	81
LOCAL AUTHORITIES AS GUARDIANS OF MONUMENTS,	81
MUSEUMS,	82
CONCLUSION,	90
APPENDIX A.—QUESTIONNAIRE OF THE COMITÉ HISTORIQUE,	96
APPENDIX B.—THE LAW AS TO INJURY TO ANCIENT MONUMENTS,	102
APPENDIX C.—THE LAW AS TO TREASURE TROVE,	104

INTRODUCTORY.

ARCHAEOLOGY A SCIENCE.

It is only within the last sixty years that Archaeology has taken rank as a science. It found no place in Whewell's *History of the Inductive Sciences*,[1] because it did not then exist. It had scarcely been recognized when *The Manual of Scientific Enquiry* was issued by the Lords Commissioners of the Admiralty in 1849, and it has not even yet been included in that valuable handbook, although in recent editions Anthropology has been added. The collecting and the study of the relics and monuments of man in past ages used to be known as "Antiquities," but the past meant the historic past, and the "antiquities" of a country was another name for its early history. The monuments of the past, it was thought, were not capable of being themselves interrogated, but as requiring to be interpreted by the facts of history.[2] Where record failed, tradition,

[1] London, 1837, 3 vols. 8vo.

[2] There were some exceptions. An instance is John Williams's *Essay on the Vitrified Forts of Scotland; An Account of some remarkable Ancient Ruins lately discovered in the Highlands and Northern Parts of Scotland* (Edinburgh, 1777, 8vo). Williams proceeded entirely upon observed facts and scientific experiment, and the question of the object

in the shape of legend, was pressed into the service, and the wildest fables were conventionally accepted as authentic. Archbishop Ussher's chronology[1] was the limit of the historic period, and the Bible narrative was regarded as the oldest and most reliable source of civil history. Under these conditions "antiquities" made little progress; antiquaries, in the absence of facts, resorted to conjecture, and were often led to propound the most ridiculous theories.

The establishment of the great Museum at Copenhagen, under the charge of Christian Thomsen, resulted in a revolution. The proper arrangement of the vast quantity of material which had been

and method of vitrifaction stands nearly as he left it, notwithstanding much subsequent writing and discussion. The evidence upon which he proceeded was insufficient, and he sometimes drew inferences where he could have observed, but he deserves credit for dealing with facts only. Fortunately there was no tradition and no current theory to mislead him. He had a clear idea of the value of ancient monuments. "By remains of antiquity we can often trace the progress or decline of the arts, the era of any practice or custom, or the time when any particular people or community flourished. And these monuments serve for ocular remembrancers that such people did flourish in such countries or neighbourhoods." *Op. laud.*, p. 72.

Twenty years earlier a suggestion had been made for the opening of cairns and barrows for the purpose of ascertaining facts bearing upon the former occupation and history of Scotland. *The Scots Magazine*, 1759, p. 462.

[1] It need scarcely be pointed out that the Bible narrative itself gives no dates. The Chronologies founded upon the Bible are merely calculations taken from the genealogies in Genesis; yet, curiously, they were accepted as authentic long after the "days" of creation had been interpreted as periods of indefinite duration. Thus Edmund Halley, while taking the human period at 6000 years, insisted on a long although definite period for the age of the Earth. He proposed to calculate it from an estimate of the rate at which the saltness of the sea increases. *Philosophical Transactions*, xxix. (1715), p. 296.

brought together was a matter of primary importance, and to this Thomsen devoted his most earnest thought and solved the problem with the most brilliant success. The secret of his success lay in his having command of a number of examples so large as to enable comparisons to be made, and in seeing that the characteristics which would be disclosed by comparison were of value.[1] Taking advantage of his opportunities, he compared the objects, noted their essential features, and so reduced them to types or classes. He thus learned that the articles of stone were intended for use, and had been used as utensils, implements, tools, and weapons of various kinds. He found a similar series of bronze articles and another of iron. Investigation showed that what were believed to be the most ancient graves contained utensils of stone only; those of later date contained bronze utensils, or bronze and stone together; the most recent contained iron. He next argued that as stone existed as material, and was everywhere at hand, its use must have preceded that of the metals, which are the product of skill and technical knowledge; and then that, as iron is still used for ordinary tools and weapons, while bronze is not, the use of iron for these purposes must have been later than that of bronze.

> Inde minutatim processit ferreus ensis,
> Vorsaque in obscenum species est falcis ahenae.[2]

---

[1] Cf. Bosanquet, *The Essentials of Logic*, pp. 143, 144 (London, 1895).
[2] Lucretius, *De Natura Rerum*, v. 1282-1294. Munro reads, *in obprobrium*, "the bronze articles went out of use." See also Ovid, *Fasti*, IV. 405, whose "*chalybeia massa latebat*" is a reproduction of Hesiod's μέλας δ'οὐκ ἔσκε σίδηρος (*Opera et Dies*, 150).

Such was the genesis, about the year 1836,[1] of the now famous "three-age-system" of the Northern antiquaries, which, subject to some qualifications and limitations, is accepted[2] as representing the general sequence of the stages in the history of human progress—the visible expression of successive phases of culture.[3] The idea was not new. It is as old as the poet-philosopher of the first century, and had become a commonplace with the antiquaries of the eighteenth century. But with them it was an assumption. With Thomsen it was a demonstration, a logical inference

[1] In 1833 Thomsen gave an outline of his suggestions in the *Mémoires de la Société des Antiquaires du Nord;* but it was only in 1836 that he published his classification in *Ledetraad til Nordisk Oldkyndighed*, translated into German the following year by C. Paulsen, and into English by the Earl of Ellesmere in 1848 as *A Guide to Northern Antiquities.*

[2] The "three-age-system" has been much and sharply criticised by German antiquaries. See Hostmann, *Studien zur vorgeschichtlichen Archäologie* (Braunschweig, 1890, 8vo). Lindenschmit, *Die Alterthümer unserer heidnischen Vorzeit*, vol. ii., Preface (Mainz, 1870). In France, M. Alexandre Bertrand maintains that the system is not applicable to France in the same unqualified manner as it is to Scandinavia. This is true, but while details vary, the general outline is the same. M. Gabriel de Mortillet, on the other hand, accepts the "three-age" system as applicable in its entirety to France; but points out that to mark the stages of culture in France it is necessary to subdivide the ages and to distinguish various epochs, which is just another method of qualifying the general proposition. The editor of *La Grande Encyclopédie* (i., p. 788, s.v. "Age") inserts articles by both, each advocating his own view. Bertrand's is reprinted in his *Archéologie Celtique et Gauloise*, p. 10 (Paris, 1889).

One of the most significant testimonies to the Northern system is the fact that M. Salomon Reinach has translated the classical work of Montelius into French (Paris, 1895, 8vo), although another French translation already existed. The system has also been recently defended by Moriz Hoernes in *Mittheilungen der anthropologischen Gesellschaft in Wien*, xxiii. (1893), p. 71.

[3] The "ages" shade into each other and overlap. The presence of stone implements does not necessarily imply that they belong to the stone

from observed facts. In this lies its value, and whatever modifications have been made or limitations placed on Thomsen's general statement, they are the result of further and closer observation. Archaeology is therefore founded upon the results of observation and inference, and has taken the place of "Antiquities," the interpretation of the past by means of written records.

Archaeology and geology, proceeding from opposite directions, have joined hands. By patient and accurate observation geologists have unrolled the story of the earth, and have placed before us a panorama of successive changes and progressive phases of development. First, we have land and sea alike without life, animal or vegetable; then, beginning with the coral polypes of the Laurentian period, succeeded by the seaweeds and zoophytes, the graptolites and trilobites of the Silurian seas, we pass onwards first to the gigantic endogens and bone-clad fishes of the coal measures, and then to the true exogenous timber trees and huge mammalia of the tertiary age, and lastly we reach the quaternary period with vegetable and animal types gradually shading into those of the present day, and ending with *homo sapiens*, man himself. Starting from the present time, the archaeologist has worked backwards far behind the historic period, and has made us acquainted with prehistoric man in two periods of development, an earlier and a

age, and to determine this other facts must be taken into consideration. The age of metals was posterior to that of stone, although in Western Europe it may have been an importation rather than a development, and it does not fit in with the *a priori* exactness that Thomsen and Worsaae demand.

later—palaeolithic man and neolithic man.¹ Some have argued that the human period extended back even into the tertiary age, and claim the existence of eolithic man.² If he did exist, he is the sole survivor of the tertiary fauna, and, as M. Reinach puts it, the hypothesis becomes a kind of scientific Noah's ark.³ As yet, however, the advocates of this earlier man have failed to establish their case. The facts on which they found, in so far as concerns Europe, are the alleged discovery, in tertiary beds, of bones of man, of bones of animals showing cuts or breaks supposed to be the work of man, and of flints supposed to show traces of intentional chipping. It is not beyond dispute whether the deposits in which the bones were found really belong to the tertiary period; but, letting this pass, the marks on the bones of animals can be accounted for by their having come in contact with a sharp edge of stone, or as having been made by the teeth of other wild animals. The human bones seem to have come from interments of later date. The flints, however, still afford ground for discussion. The three principal finds are those of Thenay near Tours and Puy-Courny near Aurillac, both in France, and Otta in the Valley of the Tagus

¹ Some maintain that there was a special Copper Age in Europe prior to the Bronze Age, and immediately subsequent to, or rather contemporary with, the later Stone Age. See Much, *Die Kupferzeit in Europa, und ihr Verhältniss zur Kultur der Indogermanen* (Wien, 1886, 8vo); Gross in *Revue d'Anthropologie*, 3ᵉ S., iii. (1888), p. 726. This, however, is disputed by many eminent archaeologists. It is principally in Hungary that the evidences of a copper age exist.

² The term proposed by De Mortillet, *Le Préhistorique*, pp. 18, 21 (Paris, 1885, 8vo); *La Grande Encyclopédie*, i., p. 798.

³ Reinach, *Antiquités nationales*, i., p. 96.

in Portugal. M. Gabriel de Mortillet maintains that these flints exhibit marks of intelligent action, but will not admit that they are the work of man. He attributes them to an intelligent precursor of man, whom he calls Anthropopithecus.[1] The only indication of artificial formation on the flints is "the bulb of percussion," but there seems little doubt that such bulbs are producible by natural causes, so that the intervention of the Anthropopithecus is not required. M. de Mortillet bases his case entirely on "reasoning founded on exact observations."[2] This, however, is just what his opponents dispute. They desiderate facts, the result of exact observation, and before the existence of tertiary man can even be provisionally admitted a very much larger volume of evidence, than is at present available, must be collected.

Archaeology, then, being founded on observation, comparison, and inference, it is essential that the material capable of being subjected to examination should be accurately ascertained and carefully protected and preserved. The historian deals with the written, the archaeologist with the unwritten or undesigned records of the past.[3] The latter

---

[1] De Mortillet, *Le Préhistorique*, p. 104; *La Grande Encyclopédie*, i., p. 798.

[2] See Bertrand, *La Gaule avant les Gaulois*, p. 27; Evans, *Ancient Stone Implements*, p. 247. Also the remarks of General Pitt-Rivers on the bulb of percussion, at the meeting of the British Association, 1894. *Nature*, l. (1894), p. 439.

[3] C. J. Thomsen styles written records, the immediate sources, monuments, the indirect sources, of history. *Guide to Northern Archaeology* p. 25 (London, 1848). Lord Bute speaks of the "documents" of history and the "monuments" of history. *Journal British Archaeological Association*, xlv., p. 2.

are no less important than the former, although long familiarity with written records and the weight that is attached to written evidence may incline us to place greater value upon them. The contents of the State Paper Office and of other public and private repositories seem enormous; but however great may be the quantity, it is after all strictly limited; and when an isolated question of history has to be examined, the evidence available is often found to be meagre or doubtful. The loss of a single membrane from a roll may destroy everything that is to be learned of a particular transaction. The importance, therefore, of cataloguing and protecting all historical records—taken in the broadest sense—has been fully recognized by the Government of this and of most other countries for many years past, and much labour and large sums of money have been expended upon this object. That it is a wise expenditure no one disputes, and the various record publications grow in favour every day.[1] While so much attention is given to written records, little has been done by Government for our unwritten records, our ancient monuments. Almost every effort in this direction that has hitherto been attempted has been by private persons or learned societies, but their work has necessarily been imperfect, and without concert, and on no definite plan or system. The number of such monuments is even more limited than the documentary, and, considering the long period to which they relate, are comparatively far fewer in number. If one

[1] Last year there were no fewer than 50,000 applications at the Record Office for access to State Papers and Records.

disappears or is destroyed, it is a loss that cannot be repaired, and hardly a week passes that we do not hear of some relic of antiquity perishing through neglect, or by mistake or design.[1] It is high time, therefore, that active steps should be taken to ascertain and register such monuments as remain, and to make due provision for their protection and preservation.

AN ARCHAEOLOGICAL SURVEY.

The first thing to be done in the interest of our ancient monuments, it seems to me, is to have an Archaeological survey of the United Kingdom made by and at the expense of Government, similar to the Topographical and Geological surveys which have already been executed. That such a survey is practicable is proved by experience. That it would be of the very greatest service to science can hardly be questioned. The Rev. Alfred C. Smith has published a valuable archaeological map of the Downs of North Wiltshire on the scale of six inches to the mile, with a relative memoir.[2] Mr. George Payne has given us a similar map of Kent, but on a much

[1] David Ure, *History of Rutherglen and East Kilbride* (Glasgow, 1793), p. 210, notes the destruction that was going on a hundred years ago. John Williams, twenty years earlier, mentions the wholesale destruction of brochs. *An Account of some remarkable ancient ruins lately discovered in the Highlands and Northern Parts of Scotland*, p. 70 (Edinburgh, 1777).

[2] *Map of one hundred square miles round Abury: shewing the British and Roman Stone and Earthworks of the Downs of North Wilts.* 1884.
Guide to the British and Roman Antiquities of the North Wiltshire Downs in a hundred square miles round Abury: being a key to the large map of the above. 2nd ed., 1885, 4to.

smaller scale,[1] and Sir John Evans another of Hertfordshire.[2] Some progress too has been made by the Congress of Archaeological Societies in union with the Society of Antiquaries in the preparation of archaeological maps of other parts of England. This is an heroic effort to overtake in part a necessary work, but it cannot take the place or have the authority of a general systematic survey, made under Government sanction,[3] and the maps are intended not for publication, but for reference in the hands of the societies for and by whom they are prepared. Apart from this, maps of this description are maps only in name. They are not the record of a survey, but are merely marked copies of sheets of the topographical survey. They are graphic representations or pictures of a rough kind, of the distribution of certain monuments, and indicate the whereabouts of certain finds. Much more is required. Each existing monument should be laid down from actual survey as exactly as a church or a lighthouse on the ordnance survey, and sites and places where finds occurred should be similarly treated.

A Government archaeological survey of certain provinces of India has been in progress for over thirty years.[4] The Archaeological Bureau of Den-

[1] *Archaeologia*, li., p. 447.

[2] *Archaeologia*, liii., p. 247.

[3] As to the advantages of a professional survey, see Cochran-Patrick in *The Athenæum*, March, 1879, p. 352; White, *Archaeological Sketches in Scotland*, i., p. 1 (Edinburgh, 1873, fol.).

The value of the work of a multitude of amateur observers depends entirely upon the editing. This necessitates a central staff of skilled men, who would be better employed in making original observations.

[4] The original survey was confined to Upper India. In 1873 the Govern-

mark has for many years been engaged in the preparation of archaeological charts of each parish, on which are indicated with precision the various objects of antiquarian interest. The Government of France in 1831, on the suggestion of Guizot, then Minister of Public Instruction, established a commission for the conservation of national monuments. In 1834, upon the initiative of M. Arcisse de Caumont, the *Société Française pour la conservation et la description des Monuments historiques* was formed for the purpose of aiding the commission and drawing up the monumental statistics of France. Under the authority of the Department of Public Instruction, the *Comité des Travaux historiques*[1] has published an archaeological repertory of France, and an archaeological dictionary of Gaul in the Celtic epoch has been begun and partly published. The *Commission de la Topographie des Gaules*, created in 1858 by the Emperor Napoleon III., and the *Commission de la Géographie*

ment of Bombay was permitted to establish a survey of its own for the Western Presidency, and somewhat later the Government of Madras likewise directed its attention to the collection of notes regarding the antiquities of Southern India and to their scientific exploration. The more advanced native states subsequently followed in the same direction. See Buhler, "Notes on Past and Future Archaeological Explorations in India," in *Journal of the Royal Asiatic Society*, for 1895, p. 649; *Catalogue of . . . Reports . . . Maps, etc., of the Indian Surveys*, p. 77 (London, 1878). An Archaeological Survey of Ceylon has also been commenced.

The Government of India and the various Local Governments have also taken active steps for the protection and preservation of archaeological objects in India. In 1881 the Office of Curator of Ancient Monuments in India was created, and Major H. H. Cole, R.E., was appointed Curator.

[1] *Le Comité des Travaux historiques et des Sociétés savantes* was instituted in 1834 on the recommendation of Guizot, and placed under the Department (*Ministère*) of Public Instruction. See *infra*, p. 50.

historique[1] have brought together a large quantity of material which has been utilized for various archaeological maps of ancient Gaul.[2] The same department is at present publishing an archaeological atlas of Tunis.[3] At the request and at the expense of M. Sigismond Dzialowski, the President of the archaeological section of the Scientific Society of Torun, M. Godfroy Ossowski prepared an elaborate archaeological map of Western Prussia, which was published with an explanatory memoir and inventory at Cracow

---

[1] Instituted under Order of 20th January, 1880, by the Minister of Public Instruction for the purpose of completing the work of the *Commission de la Topographie des Gaules*. Its function is to bring together all that relates to the historical topography of France from the earliest times to the Revolution. It is now the fifth section of the *Comité des Travaux historiques*.

[2] One of these is the *Carte de la Gaule, depuis les temps les plus reculés jusqu'à la Conquête Romaine, dressée avec le Concours des Sociétés savantes. Par la Commission spéciale instituée au Ministère de l'Instruction publique* (Paris, 1869, in 4 sheets).

M. Anthyme Saint-Paul announced his intention in 1876 of publishing a new archaeological map of France, but apparently this was never carried out. See *Congrès archéologique de France*, XLIIIᵉ session (1876, Arles), p. 371. At the Paris Exhibition of 1877 a number of prehistoric maps of various departments were exhibited, and Ernest Chantre exhibited in MS. an atlas of the prehistoric archaeology of France. *Matériaux pour l'histoire primitive de l'homme*, xiii., p. 314. The *Géographie universelle* of Elisée Reclus contains (vol. ii., p. 32) a map of prehistoric France by M. de Mortillet, but on so minute a scale as to make it a mere clue or index. In the Archaeological Congress of France at Senlis, in 1877, the Comte de Maricourt presented an archaeological map of part of the arrondissement of Senlis, with relative memoir; XLIVᵉ session, p. 22. See also *Ib.*, p. 301, as to the Department of l'Oise.

M. Cazalis de Fondouce has published an interesting note on archaeological maps with an outline of an archaeological map of the Department of Hérault in *Memoires de la Société archéologique de Montpellier*, T. vii. (1881), p. 273.

[3] *Atlas archéologique de la Tunisie*, Parts i., ii., iii. (Paris, 1892-5, fol.).

in 1881.[1] Friedrich Ohlenschlager of Munich has drawn up and published a very complete prehistoric atlas of Bavaria, with memoir and detailed lists of finds.[2] The German Government recently appointed a commission composed of military and archaeological experts to survey the Roman *limes* in Germany. The Bureau of Ethnology of the Smithsonian Institution has surveyed and reported on a great number of prehistoric monuments throughout the United States of America, and has prepared and published numerous valuable archaeological memoirs, maps, charts, plans, and plats.[3]

When the Ordnance Survey of Scotland was being organized, the Society of Antiquaries of Scotland applied, in 1855, to the Government of the day, requesting "that all remains, such as barrows, pillars, circles, and ecclesiastical and other ruins" should be noted on the survey. The suggestion was accepted, but the Secretary of State indicated that he must rely upon the Society "endeavouring on their part to assist the surveyors with local information through the co-operation of the resident gentry,

[1] *Carte archéologique de la Prusse occidentale* (*ancienne province Polonaise*) *et des parties adjacentes du Grand Duché de Posen*. [French and Polish.] fol. (Cracovie, 1881).

The Memoir consists of a Preface and Introduction, a detailed catalogue of monuments under classes, Lake dwellings, Forts, Burials by inhumation, etc., ending with Accidental Discoveries of isolated objects. This is followed by a Synoptical Table of prehistoric monuments arranged alphabetically by districts, a Table of Places, and Index.

[2] *Praehistorische Karte von Bayern*, 1879-80 (München, fol.; 15 sheets).

Tabellarische Uebersicht der Fundorte und Funde zur Prähistorischen Karte von Bayern (München, 1891, 8vo).

[3] The British Association has a committee to make an Ethnographical Survey of the United Kingdom, which has done some very useful work.

ministers, schoolmasters, and others."[1] The Society accordingly addressed a circular to the Conveners of all the counties in Scotland asking their assistance in the matter, and at the same time pressing upon the landed proprietors the necessity of doing all in their power for the preservation of ancient monuments. A considerable amount of archaeological information is accordingly to be found upon the sheets of the Ordnance Survey. Much of it seems, however, to have been collected without method or system, and to have been subjected to no criticism. It is consequently of very varying quality; sometimes it is accurate, at other times it is erroneous or misleading. The different classes of objects are not distinguished, so that the sheets are of little use for tracing their distribution. On the other hand, the positions of such existing objects as are recorded were ascertained as part of the survey, and can therefore be relied upon. But many objects are omitted; some are only partly entered, and modern works are sometimes marked as ancient. The places where finds have been made are occasionally noted, but rather roughly, and further details should have been given.

The Admiralty charts are useful to the archaeologist, so far as they go, but they necessarily deal with only a limited area of the country. They have this advantage over the Ordnance Survey, that they

---

[1] *Proceedings of the Society of Antiquaries of Scotland*, ii., pp. 102, 129. The idea seems to have originated with Mr. A. H. Rhind, who points out with much clearness the great usefulness of archaeological maps. *British Antiquities: Their present treatment and their real claims*, p. 39 (Edinburgh, 1855, 8vo).

were prepared on the responsibility of the officers engaged in the work; and the relative "Sailing Directions" often give interesting antiquarian notes collected by the surveyors. The account of the antiquities of the Long Island, for instance, is well and succinctly put;[1] and it was when engaged on the Hydrographical Survey of the Orkneys and Shetland and of the West Highlands that Captain F. W. L. Thomas collected part at least of the materials embodied in his many exhaustive contributions to the *Proceedings* of the Society of Antiquaries of Scotland and other kindred bodies.[2]

When the Ordnance Survey of Ireland was commenced in 1825, General Colby, then Director-General of the Survey, suggested that it should embrace not only a topographical survey, but also details of geology, antiquities, and other matters. The proposal was ultimately rejected on the score of expense, but it had been to some extent carried out in the survey of county Londonderry, and the preparation of a relative memoir. The memoir is illustrated with plans and sketches, and contains much useful archaeological information, and is the only thing that we have in this country of the nature of an official archaeological survey of a county.

Assuming that such a survey of the whole kingdom is to be made, by whom is it to be carried out? The case of county Londonderry is in point, and I have no doubt that the survey would be more cor-

---

[1] *Sailing Directions for the West Coast of Scotland*, Part i., p. 5 *et sqq.* (3rd ed., 1885).
[2] See Innes, *Scotland in the Middle Ages*, pp. 279, 280.

rectly, usefully, economically, and expeditiously made if put under the charge of the Board of Ordnance directly, or in conjunction with the Science and Art Department, as in the case of the Geological Survey,[1] than if entrusted to any other body. The plan followed on the topographical survey of Scotland, of referring the officers of the survey to local persons for information on archaeological matters, was a bad one. The officers were to a large extent precluded from exercising their own judgment, and the authorities on whom they were instructed to rely must have been most haphazard. The staff employed on the Archaeological survey need not be selected from specialists, or be under the direction of a specialist, and in many respects it would be better that it should not be. The required number of officers and men taken from the Royal Engineers could easily be trained to do the work. The preparation of the archaeological map of France, proposed by the *Comité historique des Arts et Monumens* in 1840, was entrusted to the military staff which was in charge of the topographical map of the country.[2] The Archaeological atlas of Tunis is founded upon the reports of the engineering staff of General Derrécagaix, who is in charge of the topographical survey of the protectorate. Acting

[1] The Geological Survey was recognized as a branch of the Ordnance Survey in 1834: it was placed under the Office of Woods and Forests in 1845, and in 1854 it was transferred to the Department of Science and Art.
 The Geological Survey of France is under charge of a Commission attached to the School of Mines, as regulated by Decree of 1st October, 1868. That of the United States is carried on by the Corps of Engineers of the United States Army, concurrently with the topographical survey.

[2] *Bulletin*, i., p. 88, 103, and elsewhere.

upon special instructions, they note and report upon all remains of antiquity, supply plans of towns and buildings, and drawings or photographs of monuments.[1] The chief of the survey of the Roman *limes* in Germany is General von Sarwey, an accomplished engineer officer. The archaeological survey of India was placed under the superintendence of Major-General Sir A. Cunningham of the Royal Engineers. One of the most valuable works upon Scottish antiquities that has been published in recent years [2] is by another officer of the same corps, Captain T. P. White of the Ordnance Survey. His two portly volumes, the result of three years' field observation in a portion of one county, are evidence of what

---

[1] The material thus collected is handed over to MM. E. Babelon, R. Cagnat, and S. Reinach, three members of the Commission on North Africa, who supplement it from the reports of the agents of the Department of Public Instruction, the narratives of travellers, and other sources, ancient and modern. (See *Matériaux*, xxi. (1887), pp. 171-204.)

The topographical sheets are taken for the archaeological charts just as they are, and consequently show all existing roads, railways, and canals, bridges, vineyards, and the like, and present boundaries. The only archaeological information delineated on the sheets are Roman roads, which are shown by broken lines. All other objects and ruins surveyed by the engineers are marked with running numbers in red, accompanied by the ancient name of the place if it is known. When the Commission have special information regarding any object, it is given in the descriptive text which accompanies each sheet under the running number of the object. The text is thus the archaeological index for reading the map.

The manifest objection to this plan is that the map itself tells nothing, but is merely a finger-post to point to an archaeological catalogue. The advantage is that it saves the expense of redrawing the topographical sheets and engraving them in a new form.

The scale of the maps is that adopted for the survey of Algeria, $\frac{1}{50000}$ (or 1.267 inches to a mile, *i.e.* our Ordnance Survey's smallest scale), and is figured in kilometres on the inner margin of each sheet.

[2] *Archaeological Sketches in Scotland*, folio, 2 vols. (Edinburgh, 1873-75).

would be forthcoming by a systematic and thorough archaeological exploration of the whole country. His material was collected when he was engaged in Argyllshire as Superintending Officer, but was not required for the purposes of the survey. This suggests that there is probably in the Ordnance Survey Office at present a considerable mass of material which could be utilized if an archaeological survey was entrusted to the department.

As preliminary to the survey, it would be desirable that provisional lists of the objects in each parish and county should be drawn up. For this purpose archaeological experts might be employed under the supervision of the Inspector of Ancient Monuments.[1]

Dr. Christison's list of the prehistoric forts in the south of Scotland is very much what is wanted as regards this class of monuments. It is far in advance of anything of the kind that has hitherto been attempted, and is an invaluable guide to these structures, and all the more so by reason of the plans and sections which he has prepared. At the same time he states that his

[1] As preliminary to a detailed Archaeological Survey of Southern India, the Government of Madras in 1881 appointed Mr. Robert Sewell of the Civil Service to draw up lists of all the known inscriptions and monumental antiquities in the Madras Presidency. These included all rude stone monuments, sculptured stones, temples, forts, and other buildings, ruins, tumuli, wells, and the like. See Sewell, *Lists of the Antiquarian Remains of the Presidency of Madras* (Madras, 1882, 4to).

The *Comité historique des Arts et Monumens* drew up a *questionnaire* or syllabus of queries as to the antiquities in each commune of France, which was distributed broadcast, and a vast amount of valuable material was in this way collected. The *questionnaire* was translated in the *Gentleman's Magazine*, xii. N.S. (1839), p. 74, and this translation is reprinted in Appendix A.

object is merely to give a general view of the subject, and to leave to others the task of investigating with minute detail and strict accuracy such of the remains as may seem worthy of greater attention.[1]

NATURE OF THE SURVEY.

Now, as to the work to be done : What is wanted is a survey of all monuments of antiquity of every kind, *e.g.* pillar stones and cromlechs, circles and alignments, cairns and barrows, camps, forts, and other earthworks, crosses, wells, churches and graveyards, crannogs, peels, castles, and other buildings, and their sites where the buildings are gone, caves, cup and ring-marked rocks, British and Celtic trackways, and Roman roads. The survey would be made as formally and precisely as if it were the survey of a boundary line or of a railway, and on a uniform method. Of certain objects photographs[2] or rubbings would be taken; in other cases drawings would be made, in some cases casts, and, where necessary, measured plans, sections, and elevations. All illus-

[1] *Proceedings of the Society of Antiquaries of Scotland*, xxix., p. 108.

[2] The most instructive view of an object is a matter requiring care and consideration, and must not be left to the photographers. " Fanatical in regard to their own craft, they almost invariably prefer a 'good negative' to a serviceable illustration ; thus, in nine cases out of ten, they choose views which supply the best light, not those which embrace, whatever the lighting may be, the most interesting features of the object which they are directed to copy."—*The Athenaeum*, November, 1870, p. 694.

See Eugene Trutat, *La Photographie appliquée à L'Archéologie* (Paris, 1879, 12mo).

Mr. J. Romilly Allen proposes a photographic survey of Scotland. *Proceedings of the Society of Antiquaries of Scotland*, xxviii., p. 150.

trations should be made on a determined system, and the same scale should be adopted, as far as possible, as regards each class of objects, and marked upon the photograph, drawing, or plan.

The field book would have to be kept as exactly as that of the land surveyor, and with as little sentiment; with the date of each observation, and the names and signatures of the officers. It must not be left to the individual observer to select such measurements and to note such particulars as he may think proper, but he ought to be provided with a schedule applicable to each class of monuments,[1] specifying (1) certain items of information, both positive and negative, that are to be recorded in all cases, (2) further information to be obtained if available, and (3) a blank in which he is to record whatever further he judges necessary. In this record nothing but observed facts would be entered.

At the same time inquiry would be made as to any local tradition, observance, and the like connected with the object, with the name, sex, age, residence, and calling of each person giving information, and, as far as may be, the authority for their statements. Specimen notes of this kind should be provided, to ensure uniformity of collection and to check irrelevancy. Every field book would of course be revised and checked by a superior officer, and the whole gone over by the office staff, and explanations given or corrections made by the field

---

[1] In the case of churches, a useful list of particulars to be recorded was prepared, for the use of Observers, by the old Ecclesiological Society and printed on sheets

workers where required, or, if necessary, a supplementary survey made.

Since the Ordnance survey of Scotland was completed, attention has been directed to the value of Place names. The survey is not altogether satisfactory in this matter. Some of the names entered are wrong, and others are doubtful. Often the most characteristic names are the Field names. These are not recorded at all, but should certainly be obtained for the Archaeological survey. The Gaelic names on the Ordnance survey are, in many instances, not in accordance with local pronunciation, and several varieties of spelling seem to have been adopted. In an Archaeological survey care should be taken to obtain the exact local pronunciation from the most trustworthy sources, and wherever possible the local spelling. The phonograph might be turned to excellent service in this matter. The local pronunciation could by this means be accurately obtained, labelled, and put away. Local dialects prevail in every part of the country, and a scientific attempt should be made to record them. A phonological museum would be of great interest, and would probably prove of advantage to students of philology. Apart from the general question, we would thus obtain reliable information as to Place names; and instead of having to rely upon the ear of one observer, we would have a phonogram that could be made to speak to any one and at any time, and to tell its own story.

After the survey has been made, the information will be transferred to the sheets of the existing topographical survey, both on the six-inch scale for accuracy and the one-inch scale for convenience; showing on the sheets the monuments surveyed, and distinguishing the several classes by different symbols, and the various epochs by different colours.[1]

[1] In Leroy's *Carte Archéologique du Département de la Seine-Inférieure*, 1859, four colours are used to distinguish different epochs, Gaulish, Roman, Frankish, and Modern; and conventional signs for the different classes of monuments. Conventional signs are also used on the archaeological map of France of 1869 (*supra*, p. 22). Mr. A. C. Smith uses six colours on his map of North Wiltshire. Mr. Payne and Sir John Evans distinguish by colour between Roman and Pre-Roman. M. Cazalis de Fondouce proposes six colours: yellow for the Palaeolithic age; blue for the Neolithic; green for the Bronze age (the colour being suggested by the patination of bronze objects); red for the Iron age; and shades of red for the epochs of that age—brownish red for the period of the Iron age anterior to the Roman epoch; vermilion for the Roman epoch, and carmine for the post-Roman periods. *Matériaux pour l'Histoire primitive de l'Homme*, xi., p. 292.

The first attempt to invent a set of international symbols was made by the Archaeological Section of the Scientific Association of Cracow, in the appointment of a commission for this object, with Count A. Przezdziecki as its president. Their report was presented to the Bologna Congress of Prehistoric Archaeology and Anthropology in 1871 (*Bologna Congress*, p. 364). At that Congress a committee was appointed to examine and report upon this paper. The committee held a meeting in Stockholm in 1874, when the question was fully discussed, and a sub-committee appointed to consider the details of the scheme. The sub-committee drew up a very interesting and instructive report, which is printed as a supplement to *Matériaux pour l'Histoire primitive et naturelle de l'Homme* (vol. x., 1875), translated into English by Professor Otis T. Mason of Columbian University, in *Report of the Smithsonian Institution*, 1875, p. 221, and also published separately. The report is abridged and described by Sir John Evans, *Journal of the Anthropological Institute* (1876), v., p. 427. See also *Matériaux*, viii., p. 353; ix., p. 316; xi., p. 291, 476; xiii., p. 314,'421; xxi., pp. 170, 172, 190.

The International Code was used in De Vesly's *Carte préhistorique de la Seine-Inférieure* of 1877 (reviewed in *Matériaux*, xiii. (1878), p. 89, 320); in Dalcan's *Carte préhistorique de la Gironde* of 1878, and in Moreau's *Carte préhistorique du département de la Mayenne*, 1878, but neither of the latter has apparently been published.

A series of symbols has been devised by the Standing Committee

How much of the detail of the present aspect of the country, as delineated on the topographical survey, should be retained on the archaeological charts is a question for consideration. Enough must in any case be given to identify the situation of the objects; but modern towns and streets, railways, canals, and the like should be indicated by light shading, so as not to distract the attention or weary the eye. They should appear simply as a haze upon the surface, the body of the map representing the physical features and the ancient monuments. It is essential that the contour and geographical features of the country be shown on archaeological maps, as these are often valuable elements in judging of the use and even of the age of particular objects, and for explaining difficulties. Thus the occupation of elevated sites may be explained by the fact that the low-lying lands were formerly thicket or marsh[1]; and on what is now dry land we may find the remains of a lake dwelling. The physical condition of the country determined its occupation, just as it caused trade to flow in certain channels.[2] The presence of a peat moss or

of the Congress of Archaeological Societies for the preparation of archaeological maps for the diagrammatic representation of antiquarian objects and sites. *Nature*, 1893, p. 272. Mr J. Romilly Allen suggests another code for his proposed archaeological survey of Wales. *Archaeologia Cambrensis*, x., 5th sec., p. 56.

In the Archaeological Atlas of Tunis conventional symbols are used to mark various objects, *e.g.* cemeteries and churches, which are varied according to circumstances; the symbol for a cemetery so as to distinguish whether it is Christian, Mussulman, or Jewish, and that for a church to show whether it is a church, a chapel, or a mosque.

[1] Smith, *Guide to the British and Roman Antiquities of the North Wiltshire Downs*, p. 43.
[2] *E.g.* The Weald of Kent. *Archaeologia*, li., p. 449.

of a raised beach or other feature may materially aid in ascertaining the age or determining the object of a prehistoric monument. The want of water in many ancient camps has often been remarked upon. The state of the country in old times may show that this was probably not the case when a particular camp was occupied. The absence of antiquities may in some cases be accounted for by the area being forest land in early times. In dealing with the Downs of North Wiltshire, Mr. A. C. Smith thought it desirable to show on his archaeological map the existing Sarsen stones, as these are a material factor in the problems connected with the monuments of the district.

As far as possible, therefore, the geology of the recent period should be transferred from the geological to the archaeological survey.[1] The sheet of the geological survey of Glasgow, on the six-inch scale, is better for archaeological purposes than the corresponding topographical sheet. When we see the land falling from the upper parts of the city to an alluvial tract along the Clyde, we understand how it was that the original occupation was on the ridge of the Rattonraw, and that the Stockwell was at first occupied by salmon fishers, and the Saltmarket by fullers and their tubs. While this is so, the map is so crowded with geological information as to the older formations, and so obscured by the features of the present city, that it

[1] Ossowski insists on the necessity of making the geology of the quaternary period the basis of an archaeological survey, at least in such a country as Western Prussia. *Carte archéologique de la Prusse occidentale*, pp. 13, 21. See also Nicolas, "Etudes préhistoriques sur la basse vallée du Rhône" in *Congrès archéologique de France*, XLIIIe session (Arles, 1876), p. 501.

is not practically serviceable for the archaeologist. A few of the spots where ancient canoes have been dug up are indicated; but if the whole of these, and the sites of other ancient remains, were accurately laid down on a fresh map, with the depths at which they were found, and if the river was shown in its original instead of its artificial condition, with its depth of water and its fords and pools, we would have some reliable archaeological material for settling the vexed question of the geology of the valleys of the Clyde and the Kelvin. Geology and archaeology are indeed the complements of each other.[1] If a question arises as to the age of, say, a find of flint implements, the geologist will deal with the evidence of situation, the archaeologist with that of form.

Passing to more recent times, the former boundaries of towns and parishes and the boundaries of parishes now absorbed in others should be given, as also other divisional areas now disused; as for instance, in Scotland, the boundaries of bishoprics, deaneries, and the like, as these throw considerable light on the ancient occupation of the country.

Along with the sheets of the archaeological survey of each district there would be issued a relative memoir containing a concise description of each

[1] Le Comte de Maricourt, in explaining his prehistoric map of part of the arrondissement of Senlis, department of Oise, divides the prehistoric period into two epochs—geological time and modern time. *Congrès archéologique de France*, XLIVe session (1877), p. 23.

Sadowski, *Die Handelsstrassen der Griechen und Römer*, deals in chap. I. with the Physiography of the country between the Oder, Vistula, Dnieper, and Niemen (Jena, 1877, 8vo.). See also the remarks of Flörschutz in his paper on the relation between Geology and Archaeology in *Annalen des Vereins für Nassauische Altertumskunde*, xxv. (1893), p. 1.

object and exact details of its size, position, and the like, and a scale-plan or section in the case of the more important, and, where necessary, a photograph and measured drawing. An archaeological index something similar to the one which accompanies Ossowski's archaeological map of West Prussia,[1] should be added, which would ultimately form the basis for an archaeological dictionary of the United Kingdom. The original survey books and other material would remain in the Public Record Office, or some other repository, for reference when required.

Such exactness in description may appear to be pedantic, but it is essential. Precise observation and accurate description are the foundation of all progress in archaeology, and one object of a general official survey is to secure a correct and impartial record of facts and so to provide the means for checking existing descriptions, and as a consequence the theories founded upon them, and for affording a more sure footing for future reasoning. Another object is to have as full and particular a description as it is possible to make, so that every one may be in the position of an observer, and that we may be able to reproduce the monument if the original be destroyed.[2]

The published accounts of even the more pro-

---

[1] An index, similar in plan, accompanies Mr. Payne's archaeological map of Kent and Sir John Evan's map of Hertfordshire, but less comprehensive and thorough.

[2] See J. Romilly Allen in *Archaeologia Cambrensis*, vii., 5th S. (1890), p. 278.

minent and best known objects often differ widely,[1] and no one is prepared to accept the measurements of his predecessors if he has an opportunity of checking them himself. The necessary details and thoroughly reliable descriptions will only be obtained by a general survey made by an official surveyor, competent to observe and record, and with no theory to support or evolve. Even with the best intentions, persons whose business is not that of surveying omit the most obvious particulars. Dr. Joseph Anderson mentions that Cairn Greg, in the parish of Monifieth, in Forfarshire, was opened in 1834 by Mr. Erskine of Linlathen, in presence of the late Lord Rutherfurd and Mr. George Dundas, advocate, and re-opened in 1864 by the late Dr. John Stuart and a number of other antiquaries; and yet neither Dr. Stuart, who wrote two separate accounts of the second examination, nor any other of those who took part in either examination of the cairn, gives the slightest hint of its size. The only information to be found on the point is in "The New Statistical Account of Forfarshire," where it is described as "a large heap of stones."[2]

There is a recognized method in every art which long experience has proved to be the best. Every

---

[1] Mr. Burgess, Archaeological Surveyor for Western and Southern India, refers to the difficulty he had in obtaining exact information from the various local persons employed to fill up schedules of preliminary information respecting the archaeological objects in their districts. Two returns often gave totally different measurements of the same object. Burgess, *Lists of the Antiquarian Remains in the Bombay Presidency* (Bombay, 1865, 4to).

[2] Anderson, *Scotland in Pagan Times—Bronze and Stone Ages*, p. 10.

land surveyor will give substantially the same account of an estate, will record the same particulars and in the same order, but hardly two archaeologists will give the same account of a hill-fort; the points taken up will be different, and the order of recording them different. From this cause, detailed accounts by competent observers often lose much of their value, as they cannot be co-related. Many drawings are worthless for want of a scale, and many plans drawn to scale are defective for want of a section. The plans and drawings of different observers are drawn to different scales, so that comparison is always difficult, and in many cases impossible.[1] Very often, too, they have been made to illustrate a particular point, not to record the facts of the monument itself. One of the advantages to be derived from an official survey would be that it would describe each monument exactly as it stands; the drawings of each class of monuments would be on the same scale, and plans and sections would be made according to a uniform rule.

Few travellers and field workers seem to have learned the use of the note book. Instead of making their description on the spot, they are too apt to take a few rough jottings and record them in their journal in the evening, or, it may be, after the lapse of a week or of a month. This leads to much inaccuracy, and figures are altered to reconcile discrepancies which would have been discovered and properly rectified if

[1] Very erroneous conclusions are sometimes drawn by comparing drawings on different scales. See, for example, a case mentioned by Sir John Evans, *Ancient Stone Implements*, p. 569.

the description had been written at the time.[1] Time, too, is generally against the non-professional observer. He is unable to note everything, and we constantly find a subsequent visitor remarking on something which has been omitted. He himself records that item of information, but disregards something else, so that it is but seldom that we have a complete record of the details of any monument. The official observer has no such difficulty. His duty is to record certain particulars in every case, and he cannot pass on to another piece of work until the first is finished.

What is material or immaterial is at present difficult to determine, and therefore everything should be set down. Taking a plain unsculptured standing stone as the simplest example, we should have information of (1) its elevation above ordnance datum, its situation and surroundings; (2) its relation to other adjacent monuments if any; (3) the kind of stone, granite, sandstone, slate, or schist, and so on; (4) its characteristics, whether it is the rock of the district or brought from a distance, and if so whence, if it can be determined; whether a boulder or quarried, whether entirely rough, or wrought, or partly wrought; if a stratified rock, whether the faces are those of the stratification; (5) its shape and dimensions; (6) whether erect or inclined, and if inclined, at what angle and in what direction; (7) its position by latitude and longitude, and the true bearing of its principal face;

---

[1] Dr. Johnson remarked, "How seldom descriptions correspond with realities; and the reason is, that the people do not write them till some time after, and then their imagination has added circumstances." Boswell, *The Journal of a Tour to the Hebrides*, p. 133 (London, 1785, 1st ed.).

(8) whatever in its appearance or situation may indicate the period when it was erected. In every case care should be taken to ascertain, if possible, whether there has been any alteration in the natural features of the spot or in the position of the object itself, *e.g.* whether there has been a change of level in the land, alteration of a water course, destruction or growth of a forest, and so on; whether the stone is in its original position or has been moved.[1] In the case of a circle or cromlech, the shape, dimensions, characteristics, and true position of every stone should be given. Whether a stone be rough or dressed may indicate whether it originally belonged to the circle or not.[2] Apparent mistakes may be of importance. The absence of a stone or stones in a circle, for instance, is to be noted, and the position of the blank ascertained, as it has been conjectured that the incompleteness of the great circles of Avebury, Stonehenge, Callernish, and others is intentional, and that it is like the imperfection of the circles and the ducts on cup-marked rocks.[3] So, too, earthworks should be accurately and minutely described, their areas calculated in every case, and an effort made to ascertain the geometrical relations which subsist between their parts.[4]

[1] Barochan Cross, Renfrewshire, for example, does not now stand in its original position, but on a spot to which it was removed last century.
[2] The importance of accuracy in minute details will not be disputed. On the subject of stone circles reference may be made to *P.S.A.*, 2nd S., x., pp. 302, 305, 306; cf. ix., pp. 141, 145, 148.
[3] Greenwell, *British Barrows*, p. 7; cf. *P.S.A.Scot.*, xxix., p. 71.
[4] The ancient earthworks in the Mississippi Valley have been evidently constructed to scale and with the most striking exactness. Wilson, *Prehistoric Man*, i., p. 271. (London, 1876).

If possible a section should be made and the character of the structure ascertained. A section of the Antonine wall proves conclusively the accuracy of its description by Julius Capitolinus as a *murus cespeticius*, a turf or feal dyke.[1] Canon Greenwell, from the nature of the material used in the formation of the barrows he examined, was able to judge of the appliances which the people who made them used and their plan of working.[2] It may be a question whether a mound is artificial or natural. When it might be either and no examination has been made, this must be noted and the entry marked as provisional. When a barrow or other monument is known to have been opened, this should be shown on the survey.

The exact orientation of every church should be ascertained.[3] Early churches were oriented to the sun, so that the light fell on the altar through the eastern door at sunrise.[4] When the practice of dedi-

---

[1] "Antoninus Pius," in the *Historiae Augustae Scriptores*, i., p. 258 (Lugd. Bat. 1671); *Monumenta Historica Britannica*, p. 65. "It is of turf upon a stone foundation, and is about four yards thick." Pinkerton, *An Enquiry into the History of Scotland preceding the reign of Malcolm III., or the year* 1056, i., p. 55.

[2] *British Barrows*, p. 5.

[3] See *Instructions de la Commission archéologique diocésaine à Poitiers*, p. 43 (Poitiers, 1851, 8vo). The Ecclesiological Society (originally the Cambridge Camden Society) prepared an Orientator for the use of its members, along with a catalogue of Saints' days.

[4] Lockyer, *Dawn of Astronomy*, p. 96; *P.S.A.*, 2nd S., xiii., p. 341. *The Athenaeum*, 5th October, 1895, p. 458. Scott, *Essay on the History of English Church Architecture*, p. 14 (London, 1881, 4to).

Durandus, *Rationale divinorum officiorum*, i., 1, § 8, says that a church is to be founded so that its head may point due east, that is, to sunrise at the equinoxes and not at the solstices. In the most ancient temples

cating churches to certain Saints was introduced, it became the rule in England, in theory at least, that the church should have its axis pointed to the rising of the sun on the Saint's day.[1] So also the true position of pillar stones, especially when they constitute a circle, should be noted, as various theories have been founded upon assumed bearings of certain stones. If it turns out that any monument has been oriented to sunrise or to a particular star, this may help to guide us to its age.

Apart from the advantages to be gained by exactitude in measurements, in correcting and establishing the present conclusions of archaeological science, it will be a material aid in making further advance. For example, a reliable register of dimensions may assist in enabling us to ascertain the unit of measurement adopted by the builders of monuments. Further, if we find one unit in use in one set of monuments and another unit in another

of pagan Rome the statue of the god placed on the altar looked to the west, so that the worshipper who prayed or worshipped looked to the east. The facade was therefore to the west, and this explains the statement of Hyginus, Frontinus, and Vitruvius, that the primitive temples of Rome were oriented to the west. Greek temples, on the other hand, were oriented to the east, that is, the facade was to the east and the worshipper prayed towards the west. This practice was adopted by Roman architects, and later Roman temples were built after this fashion. Beaudouin, "*La Limitation des fonds de terre dans ses rapports avec le droit de propriété*" in *Nouvelle Revue historique de droit français et étranger*, 1893, p. 416 (Reprinted, Paris, 1894, 8vo). Isidore, *Etymologiae*, xv., 4, § 7, says that in constructing a temple they looked towards sunrise at the equinox, but what he had in view was evidently the practice of the augurs and the agrimensors.

[1] Durandus, *The Symbolism of Churches*, by Neale and Webb, p. 21 (Leeds, 1843, 8vo). Chambers, *Divine Worship in England*, p. 1 (London, 1877, 4to).

set, presumably the builders were different, and, it may be, of widely separated ages, although there is nothing in their external appearance to indicate this. Accurate measurements, again, of buildings of historic times may afford evidence of the date when a particular unit was in use.

In passing, it may be suggested that all the objects of prehistoric times in our museums should be carefully measured and weighed. The people who produced the beautiful tools and weapons of the neolithic period, we may be certain, did not fashion them at random, but according to certain rules of proportion and to some scale, just as hammers at the present day are classed according to weight, and other tools by the width of their cutting edge. A similar inquiry as to the objects of the Bronze and Iron Ages would be equally interesting. The rough flints of the palaeolithic period do not suggest a reference to a standard, but a careful examination of these may show the prevalence of some unit, and who knows but we may thus be able to recover the normal length of the foot and thumb of the drift men and the cave dwellers!

So much for existing monuments, but the places at which portable objects have been found ought, as far as possible, to be indicated on the archaeological maps, the different classes of "finds" being distinguished by appropriate symbols.[1] The depths at

[1] On his archaeological map of Kent Mr. Payne notes discoveries of palaeolithic but not of neolithic implements. He likewise records finds of British coins and hoards of Roman coins, but not of individual examples of the latter. *Archaeologia*, li., p. 447. Sir John Evans follows the same rule as regards Hertfordshire. *Ib.*, liii., p. 246.

which they occurred, and the height of the spot above or its distance below Ordnance datum, should be shown, as also the character of the containing stratum—sand, gravel, clay, loam, or the like—and of the overlying strata, just as the depths of mines and bores with intervening strata and their characteristic fossils are shown on the geological sheets.[1]

The sites of monuments which are known to have been removed should be recorded. This should be comparatively easy when a description and particulars of the monument exist. It would be otherwise when there is no written description to refer to, and the inquiry would in that case require to be conducted with caution and judgment. If the surveying officer was informed that a tumulus had stood on an indicated spot, had been opened and certain articles removed, and the tumulus itself thereafter destroyed, it would be his duty to investigate the story and to make a report. Where the occurrence was comparatively recent, the facts would probably be ascertained. If a cairn has been swept away, he might be able to ascertain its shape and dimensions, whether it had been chambered or without chambers. If urns or other objects had been removed, information might be obtainable as to their colour and shape, and whether any, and if so, what articles had been found with them. Where any of the objects had been

---

[1] As showing the value of details in such matters, reference may be made to the case of the recent alleged discovery in the palaeolithic gravels of Kent of human remains of the palaeolithic age. *Quarterly Journal of the Geological Society*, li. (1895), p. 505.

preserved, the officer would inspect and describe, and note in whose custody they were.

In France the spots where finds have been made are reported to the *Commission de Géographie historique*, with the evidence in support of the statement. This is investigated by the Commission, and if found to be satisfactory the spot is marked on the map in accordance with the nature of the find.

Another end to be attained by an archaeological survey is to distinguish the true from the false. Objects are often regarded locally as ancient monuments which are not so. Many of the standing stones noted on the Ordnance survey are, I suspect, dropped boulders, and the eye is easily deceived by imaginary camps and serpent mounds. The Kaim of Kinprunes repeats itself, notwithstanding the sad and sorrowful example of Jonathan Oldbuck. It would be the duty of the archaeological survey officers to examine all objects which have been described or treated as ancient monuments, and, where necessary, mark them as false or as doubtful.

THE PROTECTION OF OUR MONUMENTS.

Of not less importance than a survey of the monuments of antiquity is their protection and preservation. This at present stands on a very unsatisfactory footing. Monuments as *partes soli* belong to the owner of the land, and he can do with them as he chooses. It is trespass to cross a field to look at a cairn or a ruined peel, and the proprietor can enforce the law if he thinks fit.[1] On the other hand, it is not a crime to deface or injure an ancient monument[2] unless the charge can be brought up to malicious mischief.[3] If, therefore, a tourist were thoughtlessly to overthrow and break a cross, or to obliterate an inscription, the act would probably stand on the same footing as if he were to upset a rubbing post or tear down a quack doctor's advertisement.[4] To remove a dead body from a grave is a punishable offence; to enter upon another man's property, open a tumulus and carry off a skeleton is not. It subjects a

[1] An object, however interesting it may be scientifically, is not in law a public place, and no amount of walking to it by students or other members of the public will ever entitle them to do so as of right. They can only go on sufferance. The case arose some years ago in reference to the well-known Rock and Spindle near St. Andrews. "As to the Rock and Spindle," says Lord Ardmillan, "I really think there can be very little doubt that, however interesting it may be to visitors and geologists, or however pleasant it may be to saunter along the shore in order to look at it, it is not the kind of use, nor is the place which the people visit the kind of place, which would be known in law as a public place. There is no law to support a right of sauntering, or a right of picnicing, or a right of walking out to look at the Rock and Spindle and walking back again." Duncan v. Lees, 13th December, 1870, 9 *Macpherson*, p. 274.

[2] "Nothing exposes a ruin to wanton ill-usage so much as ignorance of its history, except, indeed, in the case of those who ought to know better, and who, if not deterred by authority, remove or chip off parts of a monument because it possesses historical or artistic traditions." Cole, *First Report of the Curator of Ancient Monuments in India*, p. 35 (Simla, 1882, 8vo).

[3] A statement of the law applicable to injuries to monuments in the United Kingdom and in France and Germany is given in Appendix B.

[4] About 1853 or 1854 Mr. Baillie of Dochfour took civil proceedings in court against a farmer who had removed some stones from one of the Towers at Glenelg (p. 48) for dyke building, and had him compelled to replace them.

farmer to a fine of £20 or twelve months' imprisonment if he uses dynamite for catching fish in any water, public or private; but he is perfectly free, so far as the criminal law is concerned, to explode dynamite under a cromlech or to blow up a crannog.[1] It is a criminal offence to sleep in a loft or in a stackyard without the consent of the owner, or without such consent to encamp on or to light a fire near cultivated land;[2] but a gipsy family may occupy an old church on a moor or encamp in a hill-fort without fear of fine or imprisonment.

THE ANCIENT MONUMENTS PROTECTION ACT.

After long and persevering labour, Sir John Lubbock succeeded in 1882 in passing the Ancient Monuments Protection Act (45 and 46 Vict., c. 73); and somewhat earlier the late Mr. A. H. Rhind induced the Treasury to furbish up, and arm the authorities with, that ancient blunderbuss, the law of treasure trove.

The Ancient Monuments Protection Act is a definite piece of legislation, and is valuable so far as it goes.[3] It is, however, limited in its operation to 69 monuments in Great Britain and Ireland (29 in England, 21 in Scotland, and 19 in Ireland) specified in a schedule, and to such other similar

---

[1] 40 and 41 Vict., c. 65 (1877). This Act extends to the United Kingdom as regards public waters, including the sea. By 41 and 42 Vict., c. 39, § 12, it is extended to private waters, but in England only.

[2] The Trespass (Scotland) Act, 1865, 28 and 29 Vict., c. 56.

[3] It is framed very much on the lines of the Public Monuments Act of 1854, 17 and 18 Vict., c. 33. See Appendix B.

monuments as may from time to time be brought under it. The owner of any of the scheduled monuments may by deed constitute the Commissioners of Works the guardians of such monument, and in that case they are bound to maintain it out of such moneys as the Treasury may provide.[1] Any one may likewise, by deed or will, give or bequeath to the Commissioners of Works the estate or interest he has in any ancient monument, and the Commissioners may accept such gift or bequest if they think it expedient.

The Scottish list includes the Catstane, Ring of Brogar, the Stones of Callernish, the Towers of Glenelg, the Dun of Dornadilla, and others of a like character.

The Treasury are authorized to appoint an Inspector of Ancient Monuments, an office which is admirably filled by General Pitt-Rivers. The Commissioners have power, with consent of the Treasury, to purchase any ancient monument, and for this purpose the Lands Clauses Acts are incorporated with the Act; but, so far as I am aware, the power has never been exercised.

The Act constitutes it a punishable offence to injure or deface any of the scheduled monuments or any similar monuments of which the Commissioners may consent to become guardians, and makes provision for the prosecution of offenders.

---

[1] The guardianship thus constituted is, however, not binding upon a succeeding owner, unless (a) he derives his title from the owner who has granted the guardianship deed, or (b) he is a liferenter or heir of entail in possession, or (c) he has a power of sale.

In Ireland, which as a rule has far more attention from Government than either England or Scotland, a large number of ancient buildings are under the charge of the public authorities in virtue of the provisions of the Irish Church Act of 1869 (32 and 33 Vict., c. 42). Section 25 of that Act provided that such ecclesiastical buildings as were no longer in use, and which the Irish Church Temporalities Commissioners considered to be deserving of being maintained as national monuments, should be transferred to the Commissioners of Public Works in Ireland. By Order dated 30th October, 1880, there were accordingly vested in the Commissioners of Works 137 structures, and £50,000 for their preservation and maintenance. Extensive repairs have been effected upon most of these buildings, under the superintendence of an architect appointed by the Board, and they are now, as far as may be, in good order. They are periodically reported on, and the reports—in some cases accompanied by interesting plans and drawings—are annually laid before Parliament.

Under section 10 of the Act of 1882, Her Majesty may, by Order in Council, make additions to the list of monuments protected by the Act. This power has, however, been taken advantage of only to a very limited extent. It has been exercised on six occasions between 1887 and 1892, and 31 monuments—7 in England, 17 in Scotland, and 7 in Ireland—have been brought under the Act. The Government have, in fact, rendered the Act inoperative, as regards the future, by steadily declining to accept further monuments even when offered to them.

Ireland has again been more fortunate. The Act of 1892 (55 and 56 Vict., c. 46) authorizes the Commissioners of Public Works in Ireland to accept the guardianship of ancient monuments generally, and to apply the surplus income of the fund vested in them under the Act of 1869 towards their maintenance. A considerable number of buildings—castles, round towers, and abbeys—have already been transferred under this Act, and a number of others are awaiting the completion of vesting orders. There are thus between 170 and 180 monuments in Ireland under public protection, as against 38 in Scotland and 36 in England.

THE PROTECTION OF MONUMENTS IN OTHER COUNTRIES.

The protection of ancient monuments has long been a matter of public concern in France, but was put upon its present footing about sixty years ago. In 1834 the *Comité des Travaux historiques* was established by royal decree on the recommendation of Guizot, then Minister of Public Instruction.[1] Its function was to collect and publish all the important unpublished material for the history of France. Next year Guizot nominated a second committee under the Ministry of Public Instruction to take charge of the unpublished

---

[1] The *Annuaire des Sociétés savantes* and the *Revue des Sociétés savantes* record the labours of the committee and everything that is of interest to archaeologists; while the great work of De Lasteyrie and Pontalis, in course of publication, *Bibliographie générale des Travaux historiques et archéologiques* (Paris, 1888-93, 4to), brings together the titles of all papers published by the Archaeological Societies of France.

monuments of science and art as related to the history of France. This committee was composed of eminent men skilled in the subjects it had to deal with, and was divided into two sections—the one occupied with the sciences, the other with the arts. Guizot was unable to complete his scheme during his term of office, but his successor, M. de Salvandy, Minister of Public Instruction, in 1837 converted the sub-committee on arts into a committee of arts and monuments, practically a sub-committee on Archaeology. This sub-committee was, in fact, the suggestion of M. Vitet, then Inspector-General of Historic Monuments. The constitution of the committee has been somewhat modified in later years, and Archaeology has been made a distinct section with a special committee, the *Commission des Monuments historiques.* It consists of ten *ex-officio* members, and such others as the Minister of Public Instruction may nominate.[1] It is an expert committee of advice, which communicates on the one hand with localities and with public bodies, local societies and individual antiquaries, and on the other hand with the Minister of Public Instruction. It prepares and revises a list of monuments having an historic or artistic interest, sees that they are not interfered with, makes suggestions for their protection or maintenance, reports on injuries and on all proposed restorations. A considerable sum of public money is annually appropriated to the maintenance, repair, and

---

[1] Its official publication is the *Bulletin archéologique,* which began in 1843, and has continued till now, with some variation from time to time in its title. In 1883 Archaeology became a separate section of the *Comité.*

acquisition of ancient monuments, and the Commission advises the Minister of Public Instruction as to its distribution.[1]

The existing law providing for the protection of ancient monuments is that of 30th March, 1887,[2] which is something similar to our Act of 1882, but is much more thorough and comprehensive. It applies both to immoveables and moveables—the equivalent of our Scottish heritable moveable. Immoveables, either so by their nature or by destination, may be classed by the Minister of Public Instruction, in whole or in part, as historic monuments, in cases where their preservation is of national interest from an historic or artistic point of view. Before a monument can be classed, the consent of the proprietor must, however, be obtained. This cannot be dispensed with when the monument belongs to a private person, and the classification may be made subject to specified conditions. Where the owner is the State, a department, a commune, a vestry, or other public body, classification may be made without consent, by means of an order of public administration. The Minister can also enforce the law of expropriation for public utility, of 3rd May, 1841, as regards monuments which have been classed; in other words, he can acquire them by compulsory purchase, a procedure similar to that of

---

[1] In 1856-65 plans and descriptions of many of the buildings repaired and taken care of by the Department of Public Instruction were published at Paris in folio.

[2] *Loi et Décrets relatifs à la conservation des Monuments historiques* (Paris, 1889, 8vo); also in Rivière, *Lois usuelles*, p. 1344 (Paris, 1888, 6th ed.).

our Lands Clauses Acts.[1] When a private owner has declined to allow a monument to be classed, the Minister can likewise acquire it compulsorily for the State.

The effect of classification is, that the monument cannot be destroyed, or be restored, repaired, or altered to any extent without the concurrence of the Department of Public Instruction. It cannot be acquired by compulsory sale by any other department, public body, or corporation until notice has been given to the Minister of Public Instruction and his report has been received. The list of monuments which had been classed up to the year 1889 fills forty-one large octavo pages, and three more are applicable to Algiers.[2]

When, in the course of any excavation in land belonging to the State, to a department, a commune, a vestry, or other public establishment, any one discovers any monument, ruin, inscription, or object of archaeological, historic, or artistic interest, the Mayor of the Commune must at once take measures for its provisional protection, and must advise the

[1] Thus, in 1887, the Minister of Public Instruction, M. Spuller, addressed a special report to the President of the Republic on behalf of the Commission on Megalithic Monuments, and prayed for a decree declaring the preservation of the megalithic monuments of the Commune of Carnac to be of public advantage. On 21st September of that year a decree of expropriation was signed, and was published in the *Journal Officiel* (the equivalent of our Gazette) expropriating the alignments of Grand-Menec, and Kermario and various tumuli, menhirs, and dolmens.

H. Nicolas has an article on the acquisition by the State of the megalithic monuments in the department of Vaucluse and neighbourhood in *Mémoires de l'Académie de Vaucluse*, xiii. (1893), p. 232.

[2] Annexed to the special edition of the *Loi, supra*, p. 52, note. The list is officially given in the *Journal Officiel*, 31 March, 1887.

Prefect of the Department. The Prefect reports to the Minister of Public Instruction, who gives final orders on the subject. If the find occurs on private property, the Mayor advises the Prefect. On a report from the Prefect, and after consultation with the Commission on Historic Monuments, the Minister of Public Instruction may acquire the site, in whole or in part, by compulsory purchase.

The law as regards moveable objects of historic or artistic value is similar to that respecting immoveables, but it does not extend to objects belonging to private persons. A list of the objects classed under this branch of the law is exhibited at the prefecture of each Department, so that the public may be made aware of them. When classed an object cannot be sold, and a sale made in violation of the law is void.

Apart from the central government, the local governments take an active interest in the ancient monuments in the Department, and in many cases give grants of money for their protection or repair. In the Department of Seine-Inférieure, for instance, there is an administrative commission,[1] created in 1818, for the archaeological and monumental service of the Department. The Prefect is *ex officio* president. In 1867 it began to publish a bulletin[2] of its proceedings, and the Ministry of Public Instruction gives it an annual grant to assist in meeting the expense of this publica-

---

[1] *Commission départementale des Antiquités de la Seine-Inférieure.* See *Bulletin*, i. (1868), Preface ; and iii. (1874), Preface.

[2] The *Procès-verbaux* (1818-66) were published in two volumes (Rouen, 1864-67, 8vo). The first volume of the *Bulletin* appeared in 1868, and has been continued since then.

tion. In many other Departments there are similar commissions.[1]

In 1834 De Caumont established at Caen the *Société Française pour la conservation et description des Monuments historiques*. It holds an annual congress at some town of importance from an archaeological point of view, and has in various ways materially aided in the preservation of ancient monuments.[2]

In 1879 a separate Commission on Megalithic Monuments was appointed by the Government, with M. Henri Martin as president.[3] It drew up an inventory of menhirs, dolmens, and the like, and is charged with the preservation of megalithic monuments and with the duty of advising as to their acquisition by the State.[4]

Belgium has a Royal Commission on Art and Archaeology charged with the care and supervision

[1] *E.g.*, *Commission historique du Département du Nord*, with its headquarters at Lille ; *Commission des Monuments et Documents historiques de la Gironde* in Bordeaux ; and there are similar commissions for the Somme, Côte-d'Or, Charente, Seine-et-Marne, Aix, Avranches, Narbonne, Lorraine, Picardy, and others.

[2] See *The Gentleman's Magazine*, xi. N.S. (1839), p. 83.

The society publishes the well-known *Bulletin monumental*, which has appeared regularly from 1834 down to the present time.

Caen is also the seat of the energetic *Société des Antiquaries de la Normandie*, whose *Mémoires* have been published annually since 1824.

[3] *Arrête* of 21st November, 1879. It is a Sub-Commission of the *Commission des Monuments historiques*. The present president is M. G. De Mortillet. This Commission has for the last two years been engaged in preparing an exhibit for the Paris Exhibition of 1900, to consist of maps, lists of monuments, and the like, *L'Anthropologie*, v. (1894), p. 738.

[4] See *supra*, p. 53, note 1.

of her ancient monuments, which proceeds much on the same lines as the French Commission.[1]

The Austrian Empire has a Central Commission for the discovery and preservation of ancient monuments, established in 1850, reconstituted and placed under the Minister for Public Worship and Instruction in 1873. It is divided into three sections, the first of which deals with objects of prehistoric times and ancient art; the second with architectural objects, paintings, drawings, and sculptures—ecclesiastical and secular—of the Middle Ages and recent times down to the end of the eighteenth century. To the third is entrusted the care of historical monuments from the most ancient times to the end of the eighteenth century. The Commission has a yearly grant for carrying on its work, and publishes an annual *Jahrbuch*.[2]

In Prussia and in most of the German States there is a similar Commission. Russia has had one since 1859. In fact, almost every country in Europe, except our own, has some Authority whose duty it is to care for and protect its ancient monuments.[3]

Under the Civil Law certain things were withdrawn from the exercise of ownership, and in this way became inalienable, and were protected against injury.

---

[1] It was established in 1835, and has published since 1862, under the authority of the Minister of the Interior, an annual *Bulletin*.

[2] *Jahrbuch*, i., p. 3 *sqq*. (Wien, 1856, 4to) : Wussow, *Die Erhaltung der Denkmäler in den Kulturstaaten der Gegenwart*, i., p. 192; ii., p. 293 (Berlin, 1884-5, 8vo., 2 vols.).

[3] Wussow, *Op. laud.* passim. As to Denmark, see Worsaae, *The Preservation of Antiquities and National Monuments in Denmark*, translated into English, *Report of the Smithsonian Institution* for 1879, p. 299 ; *P.S.A.*, 2nd S., viii., p. 57 ; *P.S.A.Scot.*, xiv., p. 348.

Amongst these were *res divini juris*, which embraced *res sacrae*, *res religiosae*, and *res sanctae*. A burial-ground, for instance, was a *locus religiosus*. It could not be alienated, and if the land on which it was situated was sold, it did not pass to the buyer. To obliterate an inscription, to throw down a statue, remove a stone or a column from a monument constituted the *crimen sepulcri violati*, which was punishable in some cases by a fine of ten pounds of gold, and in others by condemnation to the mines.[1]

FINDS; TREASURE TROVE.

According to the Civil Law, treasure that one found in his own land belonged to himself. If a person without express search, but by chance, found a treasure in land belonging to another, one half belonged to the finder, and the other half to the owner of the land. This rule, and indeed the whole principle of acquisition of treasure by occupancy (*occupatio*)[2] has been considerably modified in most European countries.

Treasure trove[3] (*thesaurus inventus*), by the law

[1] Julii Paulli, *Sententiae*, i. 21, 8 ; *Cod.*, 9, 19, 1-5 ; *Dig.*, 47, 12.

[2] Apart from "treasure" and the rules introduced by Police Acts, the law is that the finder of lost money or of any portable article has a good claim to it against all the world except the true owner. Armory *v.* Delamirie, 1 *Strange*, 505 ; 1 Smith, *L.C.* p. 343 (10th edition, 1896). See also an interesting American case, Sovern *v.* Yoran (1888), 8 *American State Reports*, 293.

[3] As to treasure trove, see Rhind, *The Law of Treasure Trove, How can it be best adapted to accomplish useful results* (Edinburgh, 1858, 8vo); Article, *On Treasure Trove*, by George Vere Irving, *Journal Brit. Archaeol. Association*, xv., p. 81 ; *Notes on Treasure Trove* by Robert Temple, Chief Justice of Honduras, *ib.*, p. 100 ; Article, *Treasure Trove*, by T. H. Baylis, Q.C., in the *Archaeological Journal*, xliii., p. 341 ; and

both of England and of Scotland, falls within the Royal prerogative as part of the *regalia minora*. This provision is essentially a revenue law, and in former times, when there was much domestic insecurity, was a valuable source of the Crown's casual revenue.[1] According to Blackstone,[2] treasure trove is where any money or coin, gold, silver, plate, or bullion is found hidden *in* the earth or other private place, the owner thereof being unknown. If found *upon* the earth or in the sea, it belongs not to the Crown, but to the finder, if no owner appears. It is the *hiding*, not the abandoning, which gives the Crown the property. The law of Scotland is practically identical with that of England in so far as regards the prerogative of the Crown to "treasure hid in the earth." It is "confiscated as caduciary, whereby the owners are presumed to relinquish or lose the same."[3]

another by Professor E. C. Clark, *ib.*, p. 350. See also an article by T. G. Faussett, *Archaeological Journal*, xxii., p. 15.

Mr. Rhind originally condemned the law of treasure trove as injurious to the interests of archaeological science; *British Antiquities*, p. 46 (Edinburgh, 1855, 8vo).

The law of treasure trove has its historian, C. A. Beck, *Programma de historia et fatis doctrinae de Thesauris*, Jenae, 1719; and has also a considerable literature duly chronicled in the *Bibliotheca Juridica* of Lipenius (Lipsiae, 1757) and the two Supplements of 1775 and 1789.

[1] Madox, *History and Antiquities of the Exchequer*, i., p. 342 (2nd ed., 1769); Blackstone, *Commentaries*, i., p. 296; Chitty, *Law of the Prerogative of the Crown*, p. 152.

[2] *Commentaries*, i., p. 295.

[3] Stair, *Institutions*, 2, 1, 5; 2, 3, 60; 3, 3, 27; Erskine, *Institute*, 2, 1, 12; *Quoniam Attachiamenta*, c. 32, *Acts of the Parliaments of Scotland*, i., p. 652. Rankine, *Law of Land Ownership*, p. 224 (3rd ed., 1891). Fountainhall has a curious and interesting note on "Treasure Trove," 3 *Br. Sup.*, p. 148.

> "Think of the old days when invading bands
> Came like a deluge, swamping men and lands;
> How natural it was that many should
> Hide their best valuables where they could.
> 'Twas so in times of the old Roman sway;
> So yesterday—and so it is to-day;—
> And all lies dead and buried in the soil,
> The soil is Caesar's—his the splendid spoil."[1]

Occultatio thesauri, the fraudulent concealment of treasure trove is not a criminal offence in Scotland.[2] In England it is, and a finder may be indicted for the concealment of treasure trove. Thus, in January, 1863, a ploughman, William Butchers, ploughed up in a field in Sussex some ancient gold vessels and rings, weighing above 11 lbs. Supposing these to be brass, he sold them for 5s. 6d., or at the rate of 6d. per lb., to one Silas Thomas, with a promise from the latter of a better price if he got more. Thomas gave the articles to his brother-in-law, Stephen Willett, who took them to Brown and Wingrove, gold refiners, Cheapside, London, who melted them and gave £529 13s. 7d. for the lot. Neither Thomas nor Willett gave anything further to Butchers, the original finder. The matter leaked

---

[1] Goethe, *Faustus*, second part, act 1, translated by Anster, p. 16 (London, 1864).

[2] Hume, *Crimes*, i., p. 63 and note. It appears, however, in the *Regiam Majestatem*, i. 1 and iv. 3, following Glanvil, i. 2 and xiv. 2. David Ure mentions a case, towards the end of the eighteenth century, where two persons were prosecuted and imprisoned until they delivered up some coins and trinkets turned up by the plough in the parish of Rutherglen. *History of Rutherglen and East Kilbride*, p. 132 (Glasgow, 1793).

As to England, see Bracton, ff. 104, 120 (Rolls ed. ii. 150, 268), Coke, 3 *Inst.* 133.

out, and Thomas and Willett were prosecuted by the Crown and convicted of concealing treasure trove, and were each fined £265, and to be imprisoned until the fine was paid.[1]

The usual procedure in England is to establish the Crown's claim by inquest of office by the coroner and a jury. It must be shown (1) that the object found is treasure; (2) that it was found hidden in the earth.[2] In the Sussex case there was no evidence that the gold vessels had been hidden; this was held to be inferred from the circumstances. In a criminal prose-

---

[1] Reg. v. Thomas and Willett, 9 Cox, *C.C.*, 376; Leigh and Cave, *C.C.*, 313 ; 12 *W.R.*, 108. A similar case was tried at Dublin in 1867 for the concealment of certain silver coins of the reigns of Elizabeth and Charles and the Commonwealth. Reg. v. Toole, 16 *W.R.*, 439 ; *I.R.* 2 *C.L.*, 36.

[2] Sir John Jervis gives (*The Office and Duties of Coroners*, p. 301, London, 1829) the form of an Inquisition by the Coroner, in the year 1735, as to certain finds of gold and silver coin found hidden in a vacant place in the wall of the old mansion house of Crowcombe, in the county of Devon, and (p. 303) of another as to certain gold and silver coin found in 1825 in the ground under the site of an ancient house in Bend Street, Cambridge. These are repeated in some subsequent editions, but dropped out in the last by Melsheimer (London, 1888).

In 1875 the Coroner of Bedford held an inquest as to ten old guineas, 1685-1746, and some other gold and silver coins found in pulling down an old farm house. There were also a number of other coins, but being copper they admittedly belonged to the finder, and were not treasure trove. *Associated Architectural Societies' Reports and Papers*, xiii. (1875), p. 42. A wall find of £20 in gold and 4 nobles in 1441 is mentioned. Bentley, *Excerpta Historica*, p. 150. The gold was granted by the King to the finder. Nothing is said as to the nobles.

In 1749 and 1750 there was a wall find in the old Hall at Pillaton in Staffordshire of the value of £15,749. Apparently none of it went to the Crown, *P.S.A.*, 2nd S., p. 260.

The jurisdiction of a coroner, to inquire of treasure that is found, is retained in The Coroners Act, 1887 (50 and 51 Vict., c. 71), § 36; Taylor, *Law of Coroners*, (London, 1896).

cution, however, it would be only reasonable that the finder should have the benefit of the doubt, and that the Crown should prove its allegation.

Thesaurus inventus, as defined by Paulus, is an ancient deposit of money (*depositio pecuniae*) of which there is now no memory, so that it has not now an owner (*dominus*).[1] This has been enlarged in English practice to cover not only gold or silver coin, but also plate or bullion. The right of the Crown is therefore limited to gold or silver bullion or coin. It extends to nothing else.[2] The great bulk of the articles found nowadays are grave furniture or *ex votos* which accompanied an interment, and therefore not of the nature of concealed treasure. If bullion or coin be found on the surface of the ground, it is not treasure trove, but belongs to the finder. An ancient Celtic torc of gold found on the

[1] *Dig.*, 41, 1, 31, § 1 ; cf. *Cod.*, 10, 15, 1 ; Grotius, *De Jure Belli ac Pacis*, 2, 8, 7. The definition of Paulus is copied by Bracton, f. 120 (Rolls ed. ii. 270), and from him by Coke, 3 *Inst.*, 132. Craig, *Jus Feudale*, i. 16, 40, also adopts it.

As to the later Roman law, see *Inst.*, 2, 1, 39. Mr. Vere Irving contends (*Journal Archaeological Association*, xv., p. 82) that *thesaurus* originally included all valuable things hoarded up, and was not confined to gold and silver. He refers to *Cod.*, x. 10, 15, founding on the word *mobilia*; but his view cannot be supported.

There is a long and interesting article on *Thesaurus* in Hofmanni *Lexicon*, s. v.

[2] The law of Denmark was the same. It was only antiquities of silver and gold found in the earth that had to be surrendered to the Crown. Worsaae, *The Antiquities of Ireland and Denmark*, p. 9 (Dublin, 1846); a reprint from the *Proceedings of the R.I.A.*, vol. iii.

How inefficient is the law of treasure trove to protect even plate is shown by the case of the Dolgelly paten and chalice of silver found in 1890, which were ineffectually claimed and afterwards sold by auction in London in 1892 for £710. *Archaeological Journal*, xlix., p. 83.

surface of a field is not treasure trove. If it formed part of a concealed treasure it would be so, but if it accompanied an interment it is exceedingly doubtful whether it could be claimed by the Crown on that ground. Coin or bullion found at the bottom of a lake or in the bed of a river is not treasure trove. It is not hidden in the earth. An ancient gold ring turned up in digging a field is not treasure trove. The presumption is that it was dropped accidentally, not hidden.[1]

While the claim of the Crown to treasure trove is thus precise and limited, an attempt has been made in Scotland to extend it so as to include every relic of antiquity of every description wherever found. There is a maxim of the law, *Quod nullius est fit Domini Regis*,[2] and this has been pressed into the service. A bronze sword, for instance, is found. It has no owner; therefore, says the Treasury, it falls to the Crown. If this rule were of universal application there would be no place for the law of treasure trove; for *thesaurus inventus*, as being *nullius*, would necessarily become the property *Domini Regis*. So too every lost article as *nullius* would belong to the Crown; nothing would belong to a finder. Nor would there be any place for *bona vacantia*.[3] They

---

[1] See opinion of Sir R. B. Finlay and Mr. George H. Blakesly on a Case stated by the Society of Antiquaries in 1892, *P.S.A.*, 2 S. xiv., p. 222. See also *Ib.* xii., p. 323.

[2] This is the feudal rendering of the maxim of the Civil Law, *Quod enim nullius est, id ratione naturali occupanti conceditur*. *Dig.*, 41, 1, 3.

[3] *Bona vacantia* (ἀκληρονόμητα, ἀδέσποτα), properly speaking, means the property of one who dies without heirs, and which the Crown takes as *ultimus haeres*. See *Dig.*, 30, 1, 96, § 1; *Cod.*, 10, 10, 4; Ulpian, *Frag.* 28, 7; *Consuetudines Feudorum*, 2, 56.

would fall to the Crown as being *nullius*. The confusion is caused by a misunderstanding. A *res nullius* does not mean a thing that once had an owner, but a thing that never had an owner.[1] The maxim, however, is of very limited scope, and it is doubtful whether it applies in Scotland to anything but land.[2] Strays and waifs fall to the Crown, but under a different rule, which again excludes the general applicability of the *nullius* maxim. It is not contended on behalf of the Crown that a cairn or a tumulus falls under the extended claim; on the contrary, it is conceded that it belongs to the proprietor of the land as *pars soli*. But a claim to its contents is advanced. For this there can be no foundation. If it be admitted that the tumulus passes with the land, its contents must pass likewise. They can be got at only by destroying the tumulus. Each stone of a cairn is by itself a chattel, but the structure is *pars soli*, and so must be the urn it covers. A single burial urn embedded in the earth would equally pass as *pars soli*, just as an

[1] A *res nullius* in the Civil Law was either a thing which had never actually been appropriated by any one, or which, having been appropriated, had been intentionally abandoned. It did not include articles which had been lost or forgotten. In another and broader sense *res sacrae* were *res nullius*. *Inst.*, 2, 1, 7. See *supra*, p. 57.

[2] Erskine (*Institute*, 2, 1, 11) carefully limits the application of the rule. It really applies only in the case of land where the proprietor cannot support his title by writing. No length of possession will give him a title against the Crown, for the Sovereign, as feudal superior of all the land in the country, can only be divested of this property by grant. Erskine limits *res nullius* as respects moveables to waifs and strays. Craig (*Jus Feudale*, 1, 15, 17) is of the same opinion; indeed he had a difficulty in finding an example of a *res nullius* and selects the fish in the sea and precious stones found on the shore.

ornamental garden vase[1] or a sun-dial goes with a country house. If the new doctrine were sound, the Crown would be entitled, even against the owner of the land, to tools and weapons, flint chips and cores found on a palaeolithic floor beneath beds of soil, gravel, and clay which had been undisturbed for thousands of years. It is unnecessary to take such a claim seriously. A landowner's title covers everything in or attached to the soil; and as every grant theoretically flows from the Crown, he gets whatever at any time belonged to the Crown save anything that is *inter regalia*, such as treasure trove.

The maxim is just as inapplicable in the case of loose or portable things found on the surface. These belong to the owner of the land in virtue of possession. They are *partes soli*, and a water-worn pebble and a stone axe or a spindle whorl are equally part of the land.

A stone axe found on a road or in a public place does not fall to the Crown according to the maxim, but belongs to the finder. He must in Scotland[2]

---

[1] A prehistoric boat found in the alluvial soil of a river was held to belong to the owner of the soil, not to tenants for a term of 99 years. Elwes *v.* Brigg Gas Company, *L.R.*, 33 Ch. D., 562. See *P.S.A.*, 2nd S., xi., p. 199. The case of an absolute grant is different. There the owner parts with the land and everything in and on it. A lease is only a contract for the possession and profits of the land, and does not give the lessee right to a chattel not intended to pass. It is of no importance that the lessor was unaware of the presence of the chattel; it was not the bargain that the tenant should have it, and therefore it remains with the lessor.

[2] In terms of the General Police Act, 55 and 56 Vict., c. 55, § 412. A similar regulation applies in many local Acts both in England and Scotland.

The rule in France is the same. A circular of the Ministry of

report the find to the Chief Constable, but unless claimed by the loser within a year, the object is handed back to the finder. The statute recognizes, it does not create, the right of the finder.

The absurdity of the law as at present acted on in Scotland is illustrated by a recent case. In 1880 Mr. John Sturrock, a Fellow of the Society of Antiquaries of Scotland, obtained the consent of the tenant of the farm of Balcalk, in the parish of Tealing, to open a mound on the farm, which is the property of the Earl of Home. In the mound he found a skeleton, behind the right shoulder of which was an earthenware urn. Around the neck of the skeleton was a necklace of jet or cannel coal beads and plates, 147 in number. These articles he removed and exhibited to the Society of Antiquaries on 12th April, 1880, and subsequently discussed them in a paper, which was published in ordinary course in the *Proceedings* of the Society. The articles remained on exhibition in the Museum until 1882, when they were at his request returned to him. Mr. Sturrock had formed a valuable collection of antiquities, and on his death, in 1888, his executors in the next year advertised them for sale by auction in Edinburgh. No sooner, however, was this announced than the Earl of Home applied to the Court of Session for an interdict against the sale of the articles found on

Finance of 3rd August, 1825, directs that the finder of lost property shall deposit it in the Registry (*greffe*) of the Civil Tribunal—in Paris, at the Prefecture of Police—within twenty-four hours. If it is not claimed within three years, the general law of prescription applies (*Code Civil*, 2279); the article becomes the property of the finder, and is handed over to him, less the Court expenses for keeping it.

his land, which he then claimed for the first time as his property. This the executors disputed; alleged that the interment must have occurred at least 2000 years ago; and called upon his Lordship to produce a title as executor or next of kin of the deceased owner. When the case came before the Court on 22nd October, 1889, the Judge disallowed the claim of the Earl of Home, but indicated that if the articles were of antiquarian interest they should be claimed by the Crown. Thus prompted, the Queen's Remembrancer stepped in and claimed the objects as Crown property. The executors were not disposed to litigate with the Crown, and for the sake of peace submitted to the demand. The Crown next waived its rights in favour of the Earl of Home, on condition that he presented the articles to the National Museum, which he did.[1]

A grosser case of oppression under the guise of law it would be hard to figure. The articles clearly were not treasure. Cannel coal beads may have an antiquarian interest and a value in consequence, but this does not make them gold or silver. Lord Home based his claim on the fact that the articles were not of the nature of treasure trove, but had been taken by Mr. Sturrock from his property without his authority. In other words, his contention was that either as *partes soli* or as chattels not included in the lease to the tenant of the farm they belonged to him. That was certainly an intelligible claim, and it is not easy to understand how the Court got rid of it. If the claim of

[1] *Proceedings of the Society of Antiquaries of Scotland*, vol. xxv., p. 64.

the Crown was well founded it would strike at the contents of every museum in the country. The only title that any one can have to a prehistoric article is that of a finder, but according to the apocryphal doctrine of the Scottish Exchequer, every masterless thing falls to the Crown and no one can acquire any right to it. If the claim in the Sturrock case is in accordance with law, the Treasury could at any time demand from the University of Glasgow its collection of altars, inscriptions, and other Roman antiquities from the Antonine Wall; length of possession and lapse of time would be no answer, for *nullum tempus occurrit Regi.*

How partial and uncertain is the action of the Treasury is illustrated in the case of the Earl of Home himself. In 1864 his Lordship exhibited at a meeting of the Society of Antiquaries various gold ornaments found in 1834 on his estate of Douglas in Lanarkshire.[1] Why did the Queen and Lord Treasurer's Remembrancer not claim these along with the jet ornaments? I think he was well advised in not doing so, but that he should claim the one thing and not the other only demonstrates the absurdity of the present law and practice.

Regalia are divided into *majora*, which are *incommunicabilia*, and *minora*, which are *communicabilia*. The former cannot be separated from the Crown. The latter can, and have been frequently granted to subjects. There are many manors in England and baronies in Scotland the charters to which carry treasure trove, and in such cases when treasure is

[1] *P.S.A.*, 2nd S., ii., p. 401.

found it is claimed not by the Crown, but by the lord of the manor or owner of the barony.¹ Apart from every other consideration, therefore, the law of treasure trove is a singularly inefficient instrument for the protection of objects, even of gold and silver, as it is inoperative in the public interest over large areas of the country. Between July, 1859, and March, 1868, only twenty-four claims to treasure trove were made by the Treasury in England. All the finds with two exceptions were of gold or silver coins, principally silver. The one exception was a bar of silver found at Erith, in Kent, and to this the Crown's right was waived; the other was the Sussex find, which went to the melting pot.² Mr. Rhind admits that under the law as administered in Scotland prior to 1858 very little was recovered by the Crown. Very little has been recovered since.

A law so out of date,³ so limited, and so harsh in its operation as that of treasure trove should be swept

[1] A case in 1837, at the instance of the Duke of Northumberland, as lord of the barony of Wark, and as such entitled to treasure trove, is mentioned by Fenwick, *Treasure Trove in Northumberland*, p. 43 (Newcastle-upon-Tyne, 1851, 12mo). A case arose in 1892 in which the lord of a manor claimed against the Crown for treasure trove, Attorney-General *v.* Moore, *L.R.* 1893, 1 *Ch.* 676.

[2] See *Parliamentary Papers*, Nos. 496 of 1862; 131 of 1863; 297 of 1864; 385 of 1865; 354 of 1866; 508 of 1867; 465 of 1868. These returns were made upon the motion of Sir Jervoise Clarke Jervoise, who had a question with the Treasury in 1860 regarding a find of about 140 silver coins at Blendworth, near Horndean, Hants. See correspondence in *Parliamentary Paper*, No. 487 of 1861.

[3] It is a remnant of the time when wrecking was part of the law of Europe, and when the art of printing was claimed as part of the Royal Prerogative; see Brussius, *Principia Juris Feudalis*, pp. 86-88 (Edin., 1713); who fortifies himself with the opinion of

away, and something reasonable and consistent with modern ideas and the needs of archaeological science substituted. I am aware that the Treasury, as at present in right of the Crown, allows finders of treasure the metallic value of their finds, but this meets the case only to a very small extent, and the compensation is not adequate.[1] The enforcement of the law produces much irritation, and, as has just been mentioned, the right remaining in the Crown includes but a small portion of what is valuable as archaeological material. "It does seem to me," says the late Mr. J. F. Nicholls of Bristol, "that we shall never be safe to know all, or even one half, of the most valuable hoards that are brought to light until this law of treasure trove is abrogated."[2]

In India the law has been placed upon a clear and explicit footing.[3] Treasure is defined as "anything of any value hidden in the soil or in anything affixed thereto." Whenever any treasure exceeding in amount or value ten rupees (something less than a pound sterling) is found, the finder, whether the object has

Ahasuerus Fritsch in his *De abusibus Typographiae* and other similar but antiquated authors. Fritsch (1629-1701) treats the same subject at greater length in his *Tractatus de Typographis, Bibliopolis, Chartariis et Bibliopegis* (Jenae, 1675, 4to), translated into German by von Sincero, most probably the industrious Georg Jacob Scwindel, 1684-1752, who amongst other subjects wrote on bibliography (Regensburg, 1750, 4to).

[1] See Sir John Evans in *P.S.A.*, 2nd S., xi., p. 379; *Numismatic Chronicle*, 3rd S., vi., p. 176.

[2] *P.S.A.*, 2nd S., viii., p. 388.

[3] The Indian Treasure Trove Act, 1878 (Act No. VI. of 1878). Laws on the subject had been passed in Bengal in 1817, for Madras in 1832 and 1838, and for the Panjáb in 1872. The Act of 1878 applies, however, to the whole of British India.

been found in his own or in another's land, is bound to give notice to the Collector and to deposit the treasure with the nearest Government treasury, or give security for its production. The Collector thereupon publishes a notification requiring all persons interested to appear before him at a specified time. By non-appearance any claimant forfeits whatever right he may have to the treasure. The Collector, at his official inquiry, endeavours to ascertain amongst other things "as far as possible the person by whom, and the circumstances under which, such treasure was hidden." If the Collector, as the result of the inquiry, has reason to believe that the treasure had been hidden within one hundred years by a claimant or by one whom he represents, the Collector may adjourn the inquiry to allow such claimant to bring an action in the Civil Court. If he establishes his claim, he is entitled to the treasure; if he fails, it remains with the Collector. If at the inquiry no one satisfies the Collector that he has a *prima facie* claim to the treasure, he may declare it to be ownerless. On this declaration the Collector delivers it to the finder, or, if he was not the owner of the place where it was found, he divides it between the owner of the place and the finder, in accordance with rules fixed by the Act.

By section 10 of the Act it is provided that the Collector may at any time after making a declaration that a treasure is ownerless, and before delivering it to the finder or dividing it between the finder and the owner of the land, declare his intention to acquire it on behalf of the Government. In that case the Collector pays a sum equal to the value of the materials, with one fifth of such value in addition, whereupon it becomes the property of the Government.

It is unnecessary for the present purpose to refer to the laws of continental countries, but a short statement of some of them will be found in the Appendix.[1]

PRESERVATION OF ANCIENT MONUMENTS.

Assuming that we have an Archaeological survey, and that we are thus put in possession of an authentic inventory of the remaining monuments of antiquity throughout the country, I would suggest that separate lists be prepared for each county, according to its parishes, burghs, or other units of local government, showing the particular monuments within each area, their situation and the names of their owners or reputed owners. This list I would transmit to each County Council, and have it published and circulated in each parish and burgh in the county in so far as applicable. At the same time, I would send a copy of the entry or entries relating to individual monuments to their owner or reputed owner, and enact that this should be notice to him and his successors of the existence of such ancient monuments upon his land.

I would continue the provisions of the Ancient Monuments Protection Act, and give greater facilities for bringing it into operation. The Act would no doubt require to be modified in some respects to

[1] See Appendix C.

bring it into conformity with the general scheme I propose, but that is matter of detail which it is unnecessary to discuss.[1]

As regards monuments not under the Act, I would leave the property in these vested in the owners as at present, but I would propose to enact that the proprietor should be required, before executing or authorizing the execution of any work that might tend to injure, disturb, or alter any monument in the list, to give written notice to the Local Authority, and to do nothing until the Local Authority had an opportunity of examining and considering what was proposed. If satisfied that no substantial injury would ensue, the operation would be sanctioned. If, on the contrary, the Local Authority was of opinion that material injury would ensue, it should be entitled to forbid the work, and to make compensation to the proprietor, on payment of which the monument and its site would vest in the Local Authority on behalf of the public. A provision should likewise be inserted in the Lands Clauses Acts, Railway Clauses Acts, and other similar statutes, that nothing therein contained should authorize the interference with any ancient monuments except with

---

[1] Now that a liferenter can in England dispose of a settled estate, including the ancient monuments situated upon the property, it is rather absurd that he cannot make an effectual arrangement for the preservation of these monuments.

It may be that, as tenants for life have now such facilities for the sale of settled land, the provisions of §9 of the Ancient Monuments Protection Act would be held to apply to them. So great are these facilities that recently two separate owners in fee endeavoured to persuade the Court that they were respectively only tenants for life. Bates *v.* Kesterton, *L.R.* 1896, 1 *Ch.* 159.

the sanction of the Local Authority. Much mischief has been done in the past by the execution of works which could just as easily have been laid out so as to avoid a monument that they destroyed. It is but the other day that it was proposed to run a railway through Stonehenge, or at least so close as materially to injure it. One of the duties of the Austrian Monuments Commission is to report upon new railways, roads, and other similar works, so that care may be taken that they do not obtain authority to destroy or injure ancient monuments. At present the Society of Antiquaries is the only active guardian of our ancient monuments, but it can do little more than protest and exert the private influence of its members.

Compensation no doubt sounds formidable, but it would not be found to be so in fact.[1] At present in Scotland an annual sum of about £48,000 is available as residue grant under the Customs and Excise Act of 1890 for technical education, which is distributed amongst counties and burghs on a basis of population and valuation. A modest percentage of this grant would, as a rule, meet the whole charge; or a small rate might be levied under the Museums Acts. If a case arose in which a Local Authority was unable to meet a compensation claim, or where it was unreasonable that it should fall upon local funds, it would be for the consideration of the Treasury whether national

---

[1] Referring to the Archaeological Survey of India, Lord Canning says, "It will certainly cost very little in itself, and will commit the Government to no future or unforeseen expense." Cunningham, *Archaeological Survey of India*, i., p. 2 (Simla, 1871, 8vo).

moneys should not, as under Sir John Lubbock's Act, be devoted to the object, and wherever a proper case was made out, a grant would no doubt be obtained, either directly or through the Science and Art Department.

As I have already explained, in addition to the funds provided locally in France, an annual sum is assigned to the Minister of Public Instruction which is applied partly in the repair and protection of ancient monuments, and partly in making compensation where monuments are acquired by the State. This vote is found anything but burdensome ; and it is not asking too much that the British Government, which lavishes money on Art and foreign antiquities, should make a similar appropriation. A considerable amount of money every year falls to the Crown as *ultimus haeres*. This, I think, might very well be carried to a separate account, and applied in the protection and acquisition of our ancient monuments, just as the unclaimed funds in Chancery have from time to time been appropriated to various objects. The claim of the Crown to treasure trove has been enforced in the interests, it is maintained, of archaeology. Nothing but mischief has come of it ; but if the Crown's profit from *bona vacantia* were appropriated to the furtherance of archaeological science, substantial benefits would accrue. The funds arising from the estates of intestates reverting to the Crown would, in fact, be sufficient not only to protect our monuments, but to defray the cost of the archaeological survey which I urge.[1]

---

[1] Large sums have from time to time fallen in. In 1895 the balance,

INJURY TO MONUMENTS TO BE PUNISHED

I would extend the penal clauses of the Ancient Monuments Act to all monuments on the new lists, and make it a punishable offence to injure or deface any of these. But as monuments such as barrows, cairns, and the like, suffer fully as much from inconsiderate meddling as from malicious mischief, I would make it trespass for any one to open such a monument without the consent of the owner,[1] and on his consent being obtained I would make it necessary to obtain a licence from the Local Authority. "Experience has shown," says the Curator of Ancient Monuments in India, "that the keenest investigators have not always had the greatest respect for the maintenance of monuments."[2] In granting a licence I would suggest that the Local Authority should have power to prescribe the nature and extent of the operations, the treatment of the soil or material disturbed, the disposal of objects recovered, and the restoration of the outward form of the monument. It should also be a condition of the licence that the investigator

in course of administration and not yet handed over to the Crown, was in England £120,417, and in Scotland £42,289.

The total cost of the survey of Western Palestine, not counting such expenses as management, printing, etc., amounted to about £18,000, spread over a period of eight years. The cost of the actual work in the field is calculated to have been about £1 per square mile. *The Survey of Palestine, Memoirs*, vol. i., pp. 20, 30 (London, 1881, 4to). This cost includes the whole triangulation and topographical survey. This is already in existence for the United Kingdom, and all that is here required is a thorough revisal from an archaeological standpoint. The area of Great Britain and Ireland is 120,928 square miles.

[1] See "Constitution of the Emperor Leo the Younger, A.D. 474." *Cod.*, 10, 15, 1; Law of Russia, Appendix C.

[2] Cole, *First Report of the Curator of Ancient Monuments in India* p. 11 (Simla, 1882, 8vo).

make and deliver to the Local Authority measured plans and drawings of the object before it was interfered with and accurate details of the operations carried on, lists of the articles found and information as to the exact spot where, and the circumstances in which, each article was found.[1] The undisturbed material throughout the country available for research is every day becoming scarcer, and the time cannot be far distant when it will be exhausted. Any interference, therefore, with an ancient monument for the purpose of investigation should be carefully regulated, and an accurate record preserved of everything that is done.[2] Much mischief has happened by the breaking up of finds, to suit some system of classification or to give parts of it to several collections. One find kept entire, with a plan, section, and model of the barrow or other monument examined, is of more service to archaeology than twenty times the number of objects distributed in various parts of one museum or scattered about half a dozen different collections.

[1] How much may turn upon an accurate record is instanced by the Cannstadt skull, which has been accepted by some as typical of the cranium of palaeolithic man, and yet it is not in the inventory of the pleistocene mammalia found at Cannstadt in 1700. It is in the Stuttgart museum, but was not noticed until 1835, and there is no record of its provenance. See Boyd Dawkins, *Nature*, li., p. 195.

[2] " Les fouilles pratiquées sur les points les plus intéressants ne servent trop souvent qu'à grossir les collections des amateurs d'antiquités; de vrais trésors d'observation se trouvent ainsi perdus pour la science, et il est pénible de penser que le jour n'est pas éloigné où il ne restera plus rien de ces archives vénérables, qui, seules, peuvent éclairer de quelques lueurs les secrets de notre primitive histoire."—M. Marcellin Boule in *Revue d'Anthropologie*, 3ᵉ S., iii. (1888), p. 130.

So much as regards such monuments and their contents as are *partes soli*. Turn now to the case of accidental finds of portable objects of antiquarian interest. I would propose to provide as regards these that the finder should in every case be bound to give notice to the Local Authority, subject to the penalty of forfeiture in case of his failing to do so; the forfeiture to operate *retro* to the date of finding, so as to prevent a legal title transmitting. Upon notice being received, it should be open to the Local Authority, for a certain time, to acquire the object on making reasonable compensation to the satisfaction of, say, the Sheriff in Scotland and of the County Court Judge in England, and failing their doing so it would be returned to the finder. Articles acquired by the Local Authority should not be saleable, except on cause shown, and on the written report of the Inspector of Ancient Monuments detailing the reasons for giving his consent. This would prevent anything like trafficking in antiquities to the loss of science and the annoyance of finders.

Sir John Lubbock had great difficulty in getting the Ancient Monuments Protection Act passed through Parliament, and he had to modify its original provisions to meet the objections of sundry landowners who treated it as an attack upon property. Any such feeling has, I believe, now passed away. None of the disastrous consequences predicted by the opponents of the Bill have followed; but, on the contrary, every one I think is satisfied that the Act is reasonable in itself, and is a step in the right direction.

It may at first sight seem harsh to say that a man is

not to interfere with what, in the eye of the law, is his own property. But the restrictions upon the use of every kind of property are innumerable, and are not felt because we are accustomed to them. A proprietor cannot build within a certain distance of a turnpike road; he cannot dig away his soil so as to deprive his neighbour of lateral support; he cannot send down sewage upon a lower proprietor; he cannot divert a stream; he cannot abstract the water as it flows through his own estate; he cannot pollute it; he cannot build a dovecot unless he has a certain rental; he cannot burn his heather except at certain seasons; he cannot kill his game except within certain times, or kill his fish at all except in the method allowed by law; he cannot carry a gun except by licence; he cannot grow an acre of tobacco;[1]

---

[1] Statutes, 12 Cha. II., c. 34; 15 Cha. II., c. 7, § 17; extended to Scotland by 22 Geo. III., c. 73; see also 1 and 2 Will. IV., c. 13. The preamble of the leading Act is interesting in view of the changed relationship between this country and America. "Your Majesty's loyal and obedient Subjects, the Lords and Commons in this present Parliament assembled, considering of how great Concern and Importance it is, That the Colonies and Plantations of this Kingdom in America, be defended, protected, maintained, and kept up, and that all due and possible encouragement be given unto them; and that not only in regard great and considerable Dominions and Countries have been thereby gained, and added to the Imperial Crown of this Realm, but for that the Strength and Welfare of this Kingdom do very much depend upon them, in regard of the Employment of a very considerable part of its Shipping and Seamen, and of the Vent of very great Quantities of its Native Commodities and Manufactures, and also of its Supply with several considerable Commodities, which it was wont formerly to have only from Foreigners, and at far dearer Rates: (2) And forasmuch as Tobacco is one of the main Products of several of those Plantations, and upon which their Welfare and Subsistence, and the Navigation of this Kingdom, and Vent of its Commodities thither, do much depend; and in regard it is

he cannot paint a shield on his carriage without the authority of the Lyon King at Arms, or retain it when authorized except on a licence from the Excise; he cannot convert his grain into beer or whisky if he be so minded; he cannot work his coals except in accordance with stringent rules and subject to Government inspection. In towns a proprietor cannot build a house as he will, but must submit to scores of regulations as to the materials to be used, the size of the apartments, their lighting and ventilation, and the use of the building when finished. If a man finds a knife or a bracelet he cannot retain it, but must report it to the Chief Constable. Why should this apply to modern articles and not to ancient? There is no reason. The remedy is to place ancient monuments in a position similar to that of *res religiosae* under the Civil Law; to make them inalienable, and to make it punishable to injure or deface them.

The Ancient Monuments Protection Act makes provision for the protection of the monuments to which it applies. I would authorize Local Authorities where necessary to protect all monuments against injury. The removal of a monument to a place of safety is sometimes the only protection that is pos-

found by Experience, That the Tobaccos planted in these Parts are not so good and wholesome for the Takers thereof; and that by the Planting thereof, your Majesty is deprived of a considerable Part of your Revenue, arising by Customs upon Imported Tobacco; Do most humbly pray, that it be enacted by your Majesty," that no person after 1st January, 1660, shall set or plant any tobacco. The enactment is still law, although the "considerable dominions and countries" for whose benefit it was made, have long since ceased to belong to the Imperial Crown.

sible. The Local Authority should have such power of removal, but it should be made to a place as near the original site as possible; and should be permitted only in case of urgent necessity.¹

When a monument cannot be removed, and is in danger of perishing, provision should be made for taking a cast of it. In most cases the expense would be comparatively small, and could easily be met by the Local Authority. The larger objects of which casts are most necessary are sculptured stones and crosses, and as these are suitable for the Science and Art Department, arrangements might be made with it to defray part of the cost; casts might also be taken, subject to the approval of the Inspector General, in cases where no danger was apprehended to the object. A portion of the expense would, in some cases, be recouped by the sale of a certain number of copies.²

I would not propose to give a definition in the Act of Parliament of an article of "antiquarian interest" or of an "ancient monument," but this is not beyond the skill of the draftsman if it be required.³

---

[1] See General Pitt-Rivers in *P.S.A.*, 2nd S., xiii., p. 175.

[2] The French Commission on megalithic monuments, finding that the sculptures on the Breton monuments are being injured by exposure, took steps to prepare careful casts of them. One set of casts is destined for the Musée de Saint Germain, another for the museum of the Department in which the stones are situated. The remainder are for sale to other museums, French or foreign. *L'Anthropologie*, i. (1890), p. 501.

[3] As to a definition, see Wussow, *Die Erhaltung der Denkmäler*, i., p. 1 *sqq.*, ii., p. 314. In the Hungarian Law of 24th May, 1881, the expression art-monument (*Kunstdenkmal*) includes every structure (*Baulichkeit*) on or under the earth which is of value as an historical or artistic monument.

RESTORATION.

There is the question of restoration. How far this is to be carried, and under what authority it is to be exercised, need not be discussed at present. I merely refer to it because it requires to be kept in view; but it is a matter of administration rather than part of a general scheme for the preservation of ancient monuments.[1]

LOCAL AUTHORITIES AS GUARDIANS OF MONUMENTS.

Exception may be taken to my proposal to make Local Authorities the guardians of ancient monuments. It will probably be suggested that a special Board should be created for the purpose. In this I do not agree. The interference of what I may call outsiders is always resented, and an Inspector sent round like a policeman would not be looked upon with favour. If the matter be left in the hands of the Local Authority, local interest will be excited, and proprietors will act much more cordially with it than with a Commissioner from London, Edin-

---

[1] Lord Canning, speaking of the Archaeological Survey of India, says, "It does not contemplate the spending of any money upon repairs and preservation. This, when done at all, should be done upon a separate and full consideration of any case which may seem to claim it." Cunningham, *Archaeological Survey of India*, vol. I., p. 2 (Simla, 1871, 8vo). As to restoration, see Cole, *First Report of the Curator of Ancient Monuments in India*, p. 35 (Simla, 1878, 8vo).

Whenever restoration is sanctioned, it should be made imperative, especially in the case of churches, that careful plans of the building should be prepared as it stood prior to its being interfered with. See *Associated Architectural Societies' Reports and Papers*, xiii. (1875), p. 37.

burgh, or Dublin.¹ I have no doubt that in most Local Authorities a Committee could be found who, with the assistance of the County Engineer, would duly administer the proposed enactment. I would not abolish the Inspector of Ancient Monuments, but I would convert him into an Inspector General, and I would, in addition, appoint a suitable number of District Inspectors. These would visit their districts, confer with the Local Authorities, and see that the provisions of the Act were being carried into effect; and I would confer power upon the Inspector General, or the Science and Art Department upon the report of the Inspector General, to insist that the duty imposed by the Act on Local Authorities was properly discharged.

MUSEUMS.

The preservation of portable objects is of the utmost importance to archaeological science, and this brings me to the question of museums. Since the establishment of the great Copenhagen Museum, the idea of national museums of antiquities has prevailed, and we have such museums at Stockholm, Christiania, Saint-Germain, Edinburgh, Dublin, and elsewhere. There is much to be said in its favour, and it has worked well and with good results. It is excellent for scientists who can take advantage of the collection,

---

¹ Mr. Faussett points out that the failure of the attempt to make the law of treasure trove subservient to the interests of archaeological science was largely owing to the introduction of the policeman as the executive hand to enforce the Treasury regulations. *Archaeological Journal*, xxii., p. 30.

but it destroys local interest, and does nothing for the general diffusion of archaeological knowledge. If there was more local interest in antiquities more would be preserved, and a far stronger national sentiment excited.[1] While many objects are lost in consequence of the law of treasure trove, far more are destroyed or neglected through ignorance. The people have no means of learning what objects of archaeological interest are like. If they knew that there was such a thing as a stone axe, and that it was worth preserving, every one that turned up would be secured.[2]

What is wanted to meet this is the establishment and encouragement of local museums. Urban authorities in England and Ireland have power under the Museum and Gymnasium Act of 1891 to "provide and maintain museums for the reception of local antiquities or other objects of interest";[3] and part of my scheme would be to confer similar authority on

[1] "The encouragement of local interest in monuments is the more essential to secure the fabrics from danger." This is said of India, but is just as applicable to the case of the United Kingdom. Cole, *First Report of the Curator of Ancient Monuments in India*, p. 12 (Simla, 1882).

"Whilst great collections in the chief capitals of the world are of incalculable importance to science, its interests are also likely to be much promoted by those local museums, still unhappily not numerous, which are devoted to the illustration of all that belongs to particular and limited districts." Chambers's *Cyclopaedia*, s.v. "Museum" (1st Ed.).

[2] "Les objets qu'elles [les cachettes] recélent, n'ayant aucune valeur aux yeux de ceux qui les trouvent, ne sont même pas recueillis ou sont laissés entre des mains qui les égarent. La plupart du temps on les donne à jouer aux enfants de la maison." M. Paul du Chattellier in *Matériaux pour l'Histoire primitive de l'Homme*, xxii. (1888), p. 534.

[3] 54 and 55 Vict., c. 22. The Act does not extend to Scotland or the administrative county of London.

County Councils, so that they could form museums, either separately or jointly as might be found suitable, that is, county museums or district museums. One of the objects of the *Société Française pour la conservation et la description des Monuments historiques* was the formation of a local museum in each department of France where one did not already exist.[1] Of the museums at such places as Caen, Nantes, Rennes, Angers, it is unnecessary to say anything. They are of inestimable value to science, and create a strong local interest in archaeology and a desire to bring everything possible to them.

All portable articles of an archaeological or historical character found within the county, and becoming the property of the Local Authority, would be deposited in the local museum.[2] Whenever an object of especial interest or exceptional characteristics turned up, the Local Authority should have the option —which I have no doubt would be exercised—of transmitting it to the National Museum[3] in exchange

---

[1] *Bulletin Monumental*, i., p. 34.

[2] Mr. T. G. Faussett, when discussing how the law of treasure trove might be turned to account, advocated the giving of local finds to local museums. It was only finds of exceptional value that he thought should be given to the National Museum. *Archaeological Journal*, xxii., p. 29.

[3] Mr. A. H. Rhind suggested that when any person entitled to excavate ancient monuments, as proprietor of the ground or as having received permission from the proprietor, intended to do so, he should give notice of his intention to the Queen's Remembrancer, and that on his doing so he should be entitled to retain whatever he found, on condition of sending a list to the Society of Antiquaries of Scotland of what was found. The Society was to be entitled to have the use of the articles for exhibition and drawings of the more fragile. Rhind, *The Law of Treasure Trove*, p. 25 (Edinburgh, 1858, 8vo).

for duplicates of more common specimens, as is done in Denmark,[1] or a cast of the object itself. When the French collection of Henry Christy was divided between the British Museum and the Musée de Saint-Germain, the latter obtained casts of the more important objects which passed to London. The Scottish National Museum is a store-house of original material, a collection of evidences; the local museums would be not only the repositories of local antiquities, but places of popular instruction, and centres for the diffusion of information on archaeological subjects.[2] For this purpose, I would propose that they should have small typical ethnographical and anthropological collections, and be well supplied with casts and electrotype copies of archaeological subjects. In the Musée de Saint-Germain there is a long series of casts of objects preserved in the museums of Mainz, Vienna, Orleans, Rouen, Besançon, and others. The success of this plan is seen at the South Kensington museum. The electrotypes of the Hildesheim treasure, the Bernay find, and other similar relics which are to be found in many museums in this country and in the United States are, for practical

---

[1] Worsaae, *The Preservation of Antiquities and National Monuments in Denmark*, Smithsonian Report for 1879, p. 301.

[2] "Un musée d'antiquités doit être, non pas seulement un bazar de curiosités, mais un dépôt public où les gens d'étude puissent trouver, au besoin, des matériaux pour déterminer le caractère d'une époque, le degré de civilisation d'une nation même, s'il est possible, pour suivre les traces des conflits entre les peuples, enfin pour remonter les cours des âges, en fixant nos origines et en précisant les éléments dont la combinaison nous a fait ce que nous sommes." H. Schuermans, in *Annales de l'Académie d'archéologie Belgique*, xxii. (1866), p. 42.

purposes, as valuable as the originals. The casts of the Ruthwell Cross and other sculptured monuments made by the Science and Art Department are more useful to the student than the originals; and by means of casts a series of monuments may be brought together for examination and comparison, which would be otherwise impossible.[1] A large quantity of original material could easily be provided for local museums without injury to the national or central collections, and this, with copies or casts of typical specimens of such articles as would be necessary to make up a good teaching museum, would be found to be attractive and valuable, if well arranged, and the exhibits properly displayed with suitable explanations attached to them. Models to scale of barrows, underground houses, cromlechs, and the like are of great interest in themselves, and are an effectual method of preserving copies of monuments which are liable to perish, or which are not readily accessible. No part of the ethnographical and archaeological sections at the World's Fair at Chicago was more instructive than the set of models of the rock dwellings of New Mexico and Arizona, the great earthworks of the Ohio Valley, and other similar remains. A careful study of a model gives, in some respects, a more vivid impression and more accurate knowledge of the monument than a visit to the monument itself.

---

[1] On this point see *P.S.A.*, 2nd S., xiii., pp. 176, 321. To illustrate the development of the Celtic Cross, the Inspector of Ancient Monuments in Britain prepared models of a large number of such crosses, all carefully prepared to the scale of two inches to the foot, and showing the details of ornamentation.

A collection of models of our sculptured stones, says General Pitt-Rivers, "would be a means of bringing together a series of developmental forms which would seem to spread a wider knowledge of the subject, and might induce many to visit the originals who might not otherwise do so. If the popular interest in ancient monuments were increased by this means, Government might perhaps be encouraged to devote more money to the preservation of the originals."[1]

The retention of antiquities in the locality where they are found would also encourage finders to disclose their finds. Many a thing is secreted, because it is known that if it be given up it will be carried off to Edinburgh, and be buried in the Remembrancer's Office or be handed over to the National Museum, where the finder will never have an opportunity of seeing it. A man will part with his find if he knows that it will be deposited in the local museum, where he can see it any time, and possibly read his name as finder on the label.[1] If our ancient monuments are to be cared for, it must be largely by means of local interest, and local interest will only be elicited if the monument is retained in the neighbourhood.

I am aware of the argument in favour of national museums. The idea is that museums should represent national areas, and that every relic of antiquity of any importance found, say, in Scotland, should be deposited in the national museum, and so of England and Ireland. There is, it is contended, a national type which can only be recovered by bringing to-

[1] *P.S.A.*, 2nd S., xiii., p. 177.

gether every available relic within the national area.[1] There is no doubt some truth in this, but the type is not determined by political boundaries. Shetland and Dumfriesshire, Berwickshire and Skye, have essentially different characteristics, although they are within the same kingdom. Because the county of Northumberland is politically in England is surely no reason why its antiquarian remains should be transported to London to form a national collection. A local museum for the county or the district would, in my judgment, do far more in the interests of archaeological science than one great national collection. Nationality in this connection is just another name for centralization. Whatever be the advantages of centralization, and it has advantages, it crushes out local interest and local effort. A scheme for bringing all the sculptured stones in Scotland to the national museum in Edinburgh was recently advocated, and referring to it General Pitt-Rivers says—" I have ascertained it to be the opinion of the great majority, not only of archaeologists, but of those who are interested in keeping up the traditions and old associations of our country places, that these monuments should, whenever practicable, be preserved on their ancient sites. . . . Although the localities from which they come might be recorded, the monuments would not impress the mind so much as when seen on their own sites in the regions that gave birth to them; not to mention the bad effect of robbing country places in the

[1] See Rhind, *The Law of Treasure Trove*, p. 11; Irving, *Journal Brit. Archaeol. Association*, xv., p. 99; Cochran-Patrick in *The Transactions of the Glasgow Archaeological Society*, i., N.S., p. 367.

interest of the towns, a process which, on other than archaeological grounds, should be avoided as much as possible."[1]

Local museums, where they exist, have often been badly cared for, but this has arisen from the want of superintendence. If placed under the charge of a public body, they would be well administered, every article would be properly registered, labelled, and displayed, and catalogues would be published. The Blackmore Museum at Salisbury is equal to any archaeological museum in the country. The archaeological collections in the Mayer Museum at Liverpool, and in the Grosvenor Museum at Chester, are admirable. One of the features of the latter are the Roman monuments and inscriptions found in the neighbourhood. Many of the local museums in France and Germany are as well arranged, as well kept, and as instructive as the national museums.

It is only necessary to turn to the free public libraries to see how admirably an institution, involving very considerable expert knowledge, can be administered by a public body composed of ordinary citizens. It is too often forgotten that the statutes which provide for the establishment of public libraries equally contemplate the establishment and maintenance of museums, and in fact the first of what are known as the Libraries Acts was really a Museum Act.[2]

[1] *P.S.A.*, 2nd S., xiii., p. 175.

[2] The Powys-Land Club, while approving of the retention of the Law of Treasure Trove, suggested that, "in order to make it appropriately applicable to antiquities," it should be provided, "not only that the objects found should be dedicated to the public by the individual finder,

Burghs, parishes, and vestries have, under the existing law, full power to establish museums and to assess for their support. County Councils and District Councils, as such, have not the like power, but district museums can, as things stand, be established by the joint action of adjoining parishes. If thought desirable, the law could easily be altered so as to confer upon County Councils the power to establish such museums, and to raise the necessary funds. The Science and Art Department is authorized to make grants towards the acquisition of grants for public museums and for the acquisition of specimens of certain descriptions. The Acts have no doubt been taken advantage of principally for the establishment of industrial museums, but they apply equally to archaeological museums, and the Act of 1891 mentions such museums in express terms. What is wanted is a new Act consolidating the existing statutes, and conferring their powers on County Councils and other excepted authorities, and extending their operation to all parts of the United Kingdom alike.

CONCLUSION.

It may be suggested that an archaeological survey would put a stop to individual effort and to the work of archaeological societies. So far from this being the

---

but also that they should remain in the district in some public museum . . . where all, including the finder, may inspect and find them." *Report of Powys-Land Club* (1874).

8 and 9 Vict., c. 43 (1845). An Act for the establishment of museums in large towns. The Bill was brought in by Mr. William Ewart, M.P. for the Dumfries Burghs.

case, a survey would stimulate inquiry and guide the labours of students and observers to the objects and in the direction most likely to lead to valuable results. The geological survey has not superseded investigation, but has put existing information in a convenient shape, and has furnished geologists with a reliable index to the geological features of the country, by means of which they can with more certainty and profit pursue their individual inquiries. Were all ancient monuments properly recorded on a map, showing at the same time the contour of the country and its physical aspect, we should probably find it was, in some respects, even more instructive than a systematically arranged museum, and the energies of archaeologists would be directed to new lines of investigation, and fresh light would be thrown upon the history of man and the progress of civilization in the past.

Government spends large sums of money every year upon the preservation and protection of our records, the reproduction of fading charters, and the publication of ancient chronicles and other written memorials of the past, but it does not regard the monuments which illustrate or supplement these records. Archaeologists have raised the veil that shrouds the first epochs of man's life upon the earth, and have given us a glimpse of prehistoric times, but Government does nothing to collect or preserve the material which is essential for such investigations. The editing and interpretation of our Runic monuments we owe to Professor George Stephens of Copenhagen. For a record of the Roman inscriptions in this country we have to look to Germany or to Canada. The collections of Hübner and M'Caul are, however, far from complete, and it is not asking too much that Government should undertake a new *Corpus Inscriptionum Britannicarum*[1] to be placed alongside the *Monumenta Historica Britannica* of the Record Commissioners. We are indebted to the enterprise of the Spalding Club for Dr. Stuart's *Sculptured Stones of Scotland.* That work might well be extended so as to include not only the numerous stones in Scotland which were omitted by the editor, or which were unknown to him, but also all similar monuments in England, Wales, and Ireland. I refer to inscriptions and sculptures because they are of the same character as written monuments, and it is surely just as important that these should be carefully collected and accurately transcribed and photographed as that we should have new editions of the Chronicles of the Picts and Scots, or of the Exchequer Rolls of Scotland. We are justly proud of our long series of national records. They are magnificently housed; they are placed in charge of a corps of experts; they are calendared; great numbers are printed; many of the more important have been reproduced. The Domesday survey is a unique and an invaluable record, and the eight-hundredth anniversary of its completion was celebrated as a national event. "Domesday Book," it has been said, "is the story of the life of free

[1] In 1843, M. Villemain, then minister of Public Instruction in France, projected the publication of a collection of Latin inscriptions, similar to that now being carried on by the Academy of Berlin. It was abandoned by his successors, but the scheme was to some extent carried out by the Academy of Inscription in 1867.

England, written by the hands of the free Englishmen themselves, of every separate county (except four), eight hundred years ago. It tells how they lived, what they did, and how they kept themselves free Englishmen." This is true, but these same men, those who went before and those who came after them, have written part of their story upon the face of the country, and it is still there to interpret and explain the great survey. The written record has been facsimiled and photozincographed, it has been translated and annotated, and commented on. But it is of quite as much importance that what still remains visibly on the surface of the land to tell of its early divisions, its cultivation and tenure, should be gathered together and recorded. This is but one subject that would be elucidated by an archaeological survey. Such a survey would cast light upon every epoch and upon every phase of life in the past, and would throw a flood of light upon the present.

The quaternary period is common ground to the geologist and the archaeologist. The evidences of man's presence, his tools and weapons, both palaeolithic and neolithic, the remains of the animals that were his contemporaries, the condition of the earth when they lived, fall within the sphere of geology as well as of archaeology. They are dealt with to a certain extent in the geological survey. But why should the systematic survey stop at this point or be limited to the requirements of geological science? The monuments which are witnesses to man's presence, his life and labour, are surely as worthy to be collected and preserved as the fossil remains of extinct fauna and flora. The monuments of the past are not indeed wholly neglected by Government, for if an object be in itself artistic, in the opinion of the Science and Art department, it has the sedulous care of that department, and no money is grudged for its protection and reproduction. The Ardagh chalice, for instance, is of this description, but a Roman altar or a centurial stone, no matter how valuable it may be historically, is passed by. Can anything be more inconsistent? Why is one feature in man's activity to be selected for special attention, to the exclusion of every other? But on the ground of art itself something more is wanted. Properly to understand and appreciate the ornamentation of a slab we must have before us, as far as possible, the stand-point of the sculptor, his surroundings and ideas. But art did not originate within the historic period, as the museums of the Science and Art department might suggest. The Cave-men of the palaeolithic age—the *chasseurs de rennes* of French archaeologists—were skilful artists, especially as sculptors and as engravers on bone, working with the sharp edges and points of flint flakes.[1] The sketch of the antlered reindeer found in the cave of Kesserloch is as graceful in its conception, and as correct in its details as the work of an animal painter of to-day. It is not a solitary example, for quite a gallery of such have been collected from the caves of Switzerland, France, and Belgium.[2] Many of the

---

[1] De Mortillet, *Le Préhistorique*, p. 414.

[2] See Reinach, *Antiquités nationales*, i., p. 174. Only one example has been found in England, at Cresswell in Derbyshire.

M. Édouard Piette of Angers, dealing with the growth of the

stone tools and weapons of the neolithic period are as finished specimens of industrial art as those of the best workshops of to-day. To limit ourselves, however, to the artistic side of man's nature will give but a partial view. We wish to know his life as a whole, his surroundings, his pursuits and manner of living, everything in fact that enables us to trace the growth and development of culture and civilization. For this purpose the undesigned and unwritten records of the past must be systematically ascertained, protected, and preserved, and, if need be, copied or reproduced. To do this effectually Government assistance is essential as a first step. It is a work that has been too long neglected, and should be no longer delayed. I would therefore urge upon every one interested in the progress of archaeological science to lose no opportunity of pressing the claims of an Archaeological survey of the United Kingdom on the immediate attention of Government. Let us at once and for ever wipe away the reproach that England is the only country in Europe that does nothing to register and protect her Ancient Monuments.

industrial arts and arts of design, terms the period during which quaternary man carved or engraved bone and other material by means of flint the *Glyptic* Age, and subdivides it into two periods, the *Equidian* and the *Cervidian*, and to each of these he assigns two subdivisions. "Notes pour servir à l'Histoire de l'Art primitif" in *L'Anthropologie*, v. (1894), p. 129. Although the neolithic men were immeasurably above the Cave-men in culture, they were far below them in the arts of design. Boyd Dawkins, *Early Man in Britain*, p. 305; Bertrand, *La Gaule avant les Gaulois*, p. 163; Reinach, *Antiquités nationales*, i., p. 169. As to the art of the Cave-men (*les chasseurs de rennes*), see Reinach *op. laud.*, p. 168, and works there referred to.

APPENDIX A.

QUESTIONS addressed by the COMITÉ HISTORIQUE DES ARTS ET MONUMENS to its Correspondents; the answers to be made with precision and returned to the Committee.

§ I. GAELIC MONUMENTS.

1. Do there exist in the (commune of A. . . .) any stones or rocks consecrated by popular superstition?
2. Are these rocks adherent to the soil, or planted in the earth by the hand of man?
3. Are these rocks of the same nature as the stones of the country? and, if not, from what place and from what distance is it supposed that they have been brought?
4. What name do they bear in the district?
5. What is their number?
6. What are their height, breadth, and thickness?
7. Are these rocks arranged in a circle?
8. Are they poised *in equilibrio*?
9. Are they grouped two and two, joined by a third, placed on them transversely so as to form either a kind of table, or else a covered alley?
10. Have any designs been remarked on these stones?
11. Have any excavations of research been made near them?
12. What has been found?
13. Are there any *tumuli* or *barrows* existing, formed by the hand of man?
14. Have they been examined?
15. What has been found?

16. Are there any trees or fountains consecrated by superstitious practices?

17. At what distance from the church?

18. Are there any caves, and have any graves been found in them?

19. Are there any traditions attached to them?

20. Have any kind of wedges or hatchets in polished stone or metal been found?

§ II. ROMAN MONUMENTS.

1. Are there to be found in the (commune of A. . . .) any fragments of an ancient road passing in the district for a Roman road, or bearing the names either of "Caesar's Way" or "*Chaussée de Brunehaut*," or any other denomination conveying the idea of its ancient importance, and of an origin more or less remote?

2. What is the direction of this road? How far can it be traced? What portion of the (commune) does it traverse?

3. What name is given to it in the district?

4. What traditions are connected with its construction?

5. What are the names of the hamlets, farms, or localities traversed by it?

6. Has there been found along these roads, particularly under crosses or amidst the foundations of any religious edifice, columns nearly similar to the mile-stones of high roads, and bearing an inscription? What can be read of this inscription?

7. Are there any regular elevations or undulations of land or earth forming an inclosure, and known under the denomination of Roman camps or Caesar's camps?

8. If a road exists, does it terminate at one of these inclosures?

9. Is there any spot to which the tradition of an ancient battle-field is attached? Is this tradition supported by any authentic facts; by a significative appellation; by any vestiges of entrenchments, or by arms, bones, graves, or other objects that have been discovered?

10. Are there found in the fields at ploughing time fragments of reddish pottery, tiles, or bricks, whole or in bits, of very fine clay and of great hardness?

11. Are any medals or coins found;—any fragments of arms,

buckles, pins in bronze with or without springs, rings, short thick clumsy keys, glass objects, little cubes of clay, red, black, white, or yellow, fit for forming mosaics; little figures of men or animals in bronze or baked clay?

12. Are there to be observed, either on the surface of the ground or after excavations have been made, fragments of ancient walls, very thick, coated with small square stones, forming a regular system of work, and intersected at various distances by layers of large flat bricks?

13. What is the form of these buildings? Are they in a straight line, or do they follow a circular or semicircular direction?

14. Are fragments of marble found,—inscriptions, coins, statues, shafts of columns, capitals, pieces of sculpture, either in stone or in bronze?

15. Have there been found in places not now consecrated to purposes of worship, coffins in stone, plaister, or baked earth; placed singly or in groups? What is their direction and the nature of the stone? What has been found within? Do they bear ornaments, figures, or inscriptions? Do they appear to have been already examined?

§ III. MONUMENTS OF THE MIDDLE AGES.

1. Does the (commune of A....) possess one or more churches?
2. Are there any isolated chapels, and subterranean chapels or crypts?
3. What are the dimensions of each church? the lengths internally? the width ditto?
4. Is it in the form of a cross?
5. Is the choir terminated externally, in a rectangular or semicircular manner? Is it surrounded by chapels? Do some of these chapels form a semicircular projection, and vaulted outside of the wall?
6. Of what materials is it constructed? Are any parts of it observed to be in small squared stones (commonly 'tufa') or are there at various intervals layers of large flat bricks?
7. In the inside are there pillars [piers] or columns [shafts]? How many ranges of them are there?

8. Are the pillars square, cylindrical, or composed of a bundle of columns?

9. Are these pillars or columns ornamented with sculptured capitals?

10. What do the sculptures of these capitals represent? Is it men or animals, or pearls in strings, or embroidered work, or foliage? Can the plants be made out to which the foliage belongs?

11. Are the bases of the columns flat or raised? Are they sculptured? Are there a kind of claws or feet at their angles?

12. Are there any statues in stone, either inside or outside the church, and especially under the doorways?

13. In the interior, are there, either against the walls or above the altars, little statues in wood or alabaster, painted or gilt, placed one over the other, and representing scenes of sacred history?

14. What is the form of the windows? Are they terminated rectangularly; with a circular or with a pointed arch ('*ogive*')?

15. How many times does their height exceed their width?

16. Are they supported laterally by columns?

17. Are they divided internally by stone separations? These separations—are they perpendicular, curved, or circular?

18. The windows—are they in white or in coloured glass? Are figures to be distinguished on them? What is the size of these figures? The colours—are they light or dark? Is the flesh of the figures represented by the white glass, or by a tint more or less brown? Do the figures come out upon a dark blue ground, or on a ground of landscape and architecture? On the glass, are there any inscriptions (*légendes*—labels bearing characters) to be distinguished, either in Latin or in French? Can they be read and copied? Is there no date to be found in these inscriptions?

19. If the walls and pillars are covered with lime or whitewash, cannot their coating be got off in some places, and are not traces of ancient paintings to be found on the stone?

20. Are the vaultings of the church circular or pointed; in wood or in stone? Are they painted or merely whitened? Do the edges of the vaulting project? Are their ribs angular or rounded? Are they terminated at their points of junction by circular keystones (*rosaces*) more or less sculptured, or by pendent sculptured stones (*culs de lampe*)?

21. Is there merely a ceiling instead of vaulting? Are the beams visible? Are they painted, sculptured, or perfectly plain?

22. Are the stalls of the choir or the pulpit sculptured; in wood or in stone?

23. Are there to be found in the church great flags of stone or marble serving for the pavement, and on which are traced figures of men or women, ecclesiastics or knights? Is the inscription which ought to surround these figures legible? Can it be copied?

24. Do there exist in the church any other kind of tombs, with or without statues, with or without inscriptions?

25. Are the doorways of the church rectangular, circular, or pointed? Are they supported by one or several ranges of columns? Are there any statues between the columns? What do the capitals of these columns represent? Have the doorways only one opening, or is there a pillar dividing them in the middle? Is there a bas-relief above the opening or openings? What does it represent? Of what size are the figures?

26. Is the church entered immediately, or is there a porch within or without the portal?

27. Is the roof of the church flat or pointed; covered with tiles, slates, or lead; surrounded with open work stone battlements (*galeries*)?

28. What is the form of the cornice or capping? Is it supported by little square stones representing the ends of beams, and terminated by figures of men and animals, commonly monstrosities, or by small arches, or by a kind of consoles or modilions (*corbels*)? Is it accompanied by trefoils or quatrefoils, hollowed out? Does the cornice or capping consist of mouldings, or of a running ornament with foliage?

29. Are the walls sustained by buttresses? Are these buttresses adherent to the wall? Are they detached from it, and do they support it by means of flying buttresses? Are they plain or ornamented with sculpture?

30. Is the church surmounted by one or more towers? On what part of the edifice are these towers placed? What is their form; round, square, or octagonal? Do they contain a staircase? Are they terminated by a platform, or by a roof or a spire? Is this roof or spire constructed of wood or stone, and covered with slates, tiles, or lead?

31. Does there exist in the (commune of A....) any ancient abbey or convent? Of what religious order, and dedicated to what saint? Are there any remains of the conventual buildings in existence? Does the cloister still remain?

32. Are there to be found at the crossways of the (commune) or in the cemetery any stone crosses? What are their dimensions? Are they ornamented with sculptures?

33. If any isolated chapels exist, are they near to any fountain (spring) frequented by the sick? Do people go thither in pilgrimage? Do these pilgrimages take place on the eve of the saint's-day or on the saint's-day itself? What local customs or peculiar ceremonials are observed there? What kind of invalids go there?

34. Is there any ancient castle in the (commune of A....)? Is it fortified? Is it in ruins or in good condition, inhabited or deserted?

35. If it is fortified, are the towers round or square, truncated above or crowned with battlements? Is it surrounded by fosses; with or without machicolations? Is there a donjon-keep? Are there any vaults?

36. What are the shape and dimensions of the windows; are they plain or decorated?

37. In the interior, are the chimney-places large? Are they ornamented with sculptures in stone, marble, or wood? Are the ceilings and wainscotings painted or sculptured? Are traces of ancient armorial bearings to be seen on the walls? Who were the proprietors before 1789 (the commencement of the great revolution)? Do the old men of the district know of any tradition relative to the castle?

38. Does any other house exist in the district ornamented with painting, sculpture, or decorations in wood or stone?

39. Is anything known, either in the castle or the church, or anywhere else, of any pictures, tapestries, ancient carved furniture, title-deeds, or archives, medals, family portraits, altar-ornaments, or, in short, any other objects belonging to an epoch more or less remote?

APPENDIX B.

THE LAW AS TO INJURY TO ANCIENT MONUMENTS.

I. THE UNITED KINGDOM.

AFTER the smashing of the Portland Vase in the British Museum, an Act, 8 and 9 Vict., c. 44 (1845), was passed for the due punishment of such offences; and the enactment is repeated in the Malicious Injury to Property Act of 1861 (24 and 25 Vict., c. 97), § 39.

After the disappearance of the statue of George I. from Leicester Square, London, and the abstraction of the sword of the statue of Charles I. at Charing Cross, an Act (17 and 18 Vict., c. 33, 1854) was passed placing certain public statues under the Commissioners of Works, making provision for their maintenance and repair, and making it punishable to injure them. The latter part of the enactment is now superseded by the Act of 1861.

These Acts do not extend to Scotland, as malicious mischief is a common law offence in this country, punishable by fine or imprisonment according to the gravity of the case.[1]

Under the English Acts, malicious injury to an object of art or the like in a museum or other public building is constituted a special offence, and by § 51 of the Act of 1861 the law is extended to malicious injury to any property where the damage exceeds £5. This covers the case of an angry man smashing a draper's plate-glass window, but who is to assess the damage to an ancient monument? For five shillings a mason might be able to make it look as well as ever, but the work of the iconoclast may have destroyed its characteristic features. The case is perhaps even more difficult to reach in Scotland, as to constitute malicious mischief there must be the indulgence of cruel or malicious passion or an attempt to concuss others by injuring their property. The law of Scotland does not distinguish between objects in a field and those in a museum. The law of England is much more stringent; objects in museums are protected, be their commercial value small or great, and the same applies to public statues and memorial monuments in churches and churchyards.

[1] Macdonald, *Treatise on the Criminal Law of Scotland*, p. 115, ed. 1894.

Defacing of monuments in churches or churchyards, says Lord Coke, is punishable by the common law.[1] When a church is rebuilt, the monuments are to be carefully taken down and re-erected in the new church (The Church Building Act, 1819, 59 Geo. III., c. 134, § 40). It is a punishable offence in England to injure or deface a monument in a cemetery (The Cemeteries Clauses Act, 1847, 10 and 11 Vict., c. 65, § 58). The statute 3 and 4 Edward VI., 10, for destroying images and pictures, spared tombs and monuments.

II. FRANCE.

"Whoever shall have destroyed, thrown down, mutilated, or defaced monuments, statues, and other objects destined for purposes of public utility or ornament, and erected by the public authority or with his sanction, shall be punished by imprisonment for a period of from one month to two years, and a fine of from one hundred to five hundred francs."[2]

III. GERMANY.

§ 303. "Any one who maliciously and unlawfully damages or destroys property belonging to another shall be punished by a fine up to 1000 shillings, or by imprisonment with labour up to two years. The attempt to commit this offence is criminal.

§ 304. "Any one who maliciously and unlawfully damages or destroys objects of veneration of a religious society existing in the State, or things dedicated to divine service, or tombs, public monuments, objects of art, science, or trade preserved in public collections or publicly exhibited, or objects which serve for public use or for the beautification of public roads, places, or pleasure-grounds, shall be punished by imprisonment up to three years, or by a fine up to 1500 shillings (marks). Together with the punishment of imprisonment with labour, the sentence may include the forfeiture of civil privileges. The attempt to commit these offences is criminal."[3]

---

[1] 3 *Inst.*, 202 : see Little, *The Law of Burials*, p. 61, 2nd ed. (London, 1894).

[2] *Code Pénal*, § 257, "Dégradation des monumens."

[3] *The Criminal Code of the German Empire.* Translated by Geoffrey Drage. §§ 303, 304, p. 278 (London, 1885).

APPENDIX C.

THE LAW AS TO TREASURE TROVE.

I. THE CIVIL LAW.

"THE Emperor Hadrian, in accordance with natural equity, granted treasure which a man found in his own land to the finder; and made a similar grant in the case of him who accidentally found a treasure in a sacred or religious place. But if a person, without express search but by chance, found a treasure in land belonging to another, he granted half to the finder and half to the owner of the land. Consequently, if anything be found in the Imperial demesne (*in Caesaris loco*), he ordained that half should go to the finder and half to the Emperor. In like manner, if a man find anything in ground belonging to the Treasury (*fiscus*) or to the public, one half will belong to the finder and the other to the Treasury or the city (*civitas*)."[1] This rule was substantially repeated in a constitution of Leo the Younger, A.D. 474.[2] As a reason for dividing between the finder and the owner of the soil, the constitution adds, "For thus it will happen that each will enjoy his own and not gape for that of his neighbour." The civil law was practically re-enacted in a constitution of the Emperor Frederic.[3]

---

[1] *Inst.*, 2, 1, 39, Orloff, *Commentatio Iuris Romani de Thesauris* (Erlangen, 1818).

[2] *Cod.*, 10, 15.

[3] *Consuetudines Feudorum*, 2, 56. See *A Summary View of the Feudal Law with the differences of the Scots Law from it*, p. 55 (Edinburgh, 1710). This is a curious and interesting *résumé*, by John Dundas, of the *Constitutiones Feudorum*, and was published anonymously. Mr. Dundas mentions in his preface that the book had been "given in to be printed by a certain person when the author was out of town, and knew nothing of the publishing of it, not having design'd so soon to send it abroad into the world." The *Consuetudines Feudorum* was compiled about the middle of the twelfth century at Milan, and represents the feudal law of Lombardy. It was of great authority in Germany, but not in England or Scotland.

II. THE PRUSSIAN LAW.

The General Prussian Code of 1794, revised in 1803,[1] contains a series of minute rules regarding hidden treasure, by which is understood everything of any value, found upon the surface of the soil or hidden underneath it, of which the owner is unknown. Notice of the find must be given and claims invited, except when the treasure has been buried for at least a hundred years. Treasure found by chance in another's land, and to which no claim is established, belongs one half to the finder and one half to the owner of the place. The rights of co-proprietors and conterminous proprietors of land in which treasure has been found, of usufructuaries, trustees, long lease-holders, and others are all dealt with, and provision is made (§ 86) for the case of the treasure seeker who uses magic, spirit raising, and the like, either for deception or through superstition. He is not only liable to the punishment specified in the Penal Code, but likewise forfeits his right to any treasure found by pure chance.[2]

[1] *Allgemeines Landrecht für die Preussischen Staaten*, Part 1., Title 9, § 74-106; ed. Schering (Berlin, 1876); and see Koch's commentary in his edition, i., p. 433 (Berlin, 1870).

[2] There are many references in Goethe's *Faust* to the use of magic for treasure seeking, and there is his beautiful ballad, *Der Schatzgräber* :

> "Arm am Beutel, krank am Herzen,
> Schleppt' ich meine langen Tage.
> . . . .
> Und, zu enden meine Schmerzen
> Ging ich einen Schatz zu graben."

The law bearing on treasure seeking by magic has a large literature, which will be found in Lipenius. Amongst others are *De Thesauro arte magica invento*, by J. B. Friese (Jenae, 1719); and *De pactis hominum cum Diabolo circa abditos in terra thesauros effodiendos*, by Michael Förtsch (Jenae, 1719, and Lipsiae, 1741).

These were published in connection with a case of treasure seeking at Jena, on Christmas-eve, 1715, by a medical student and two peasants under a compact with the devil, which caused immense excitement at the time. *Wahrhafftige Relation dessen was in der Heil Christ-nacht zwischen den 24 und 25 Decemb. 1715, allhier bey der Stadt Jena in einem den Galgen, nah-gelegenen Weinberg mit einer schandlichen Conjuration und Beschwerung des Satan an einem Studioso und zwey Bauern sich zugetrafen hat* (Jena, 1716, 12mo). On the general subject see Horst,

III. THE LAW OF AUSTRIA.

The Austrian Ordinances of 24th February and 2nd November, 1779, of 14th February, 1782, and 5th March, 1812, relate to finds of ancient coins and their transfer to the Cabinet of Coins, compensation being made. The Ordinance of 1812 extended their scope to antiquities of bronze or stone, such as figures, statues, busts, weapons, carved stones, etc., inscribed stones, and grave stones. The compensation for finds purchased by the authorities was to be on a reasonable scale.[1]

The general Austrian Code makes regulations as to all things which are found of greater value than two florins, including treasure, which is specially defined (§ 398). "If the discovered thing consists of money, jewellery, or other precious things, which have been so long concealed that their former proprietor can no longer be discovered, they are called treasure." Any ordinary article to which no effectual claim is made within one year belongs to the finder. In the case of treasure, the finder gets only one third, one third falling to the owner of the land, the remaining third to the State.[2]

IV. THE LAW OF HUNGARY.

The edicts of 6th August, 1812, 3rd August, 1813, and 11th April, 1815, relating to treasure trove, extend to portable monuments, as treasure, according to Hungarian law, includes objects of scientific and artistic interest, such as Greek, Roman, Pagan, or more recent trinkets, weapons, and ancient coins. Every find of his description must be notified to the authorities, by whom the value of the object is determined. The value is divided amongst the owner of the place where the find was made, the Treasury,

---

Zauber-Bibliothek, v., p. 143 (Mainz, 1825, 8vo); Grimm, *Teutonic Mythology*, iii., p. 970 (London, 1883).

The provision of the General Prussian Code (Part ii., Title 20, § 6) punishing the use of magic was repealed in 1871, and the Penal Code of that year substituted.

[1] Wussow, *Die Erhaltung der Denkmäler*, i., p. 192.

[2] *General Code for the Austrian Monarchy*, translated by Winiwarter, § 389-401 (Vienna, 1866, 8vo).

and the finder, or, where notification has been neglected, the informer. If the object possesses scientific or artistic value, the Hungarian National Museum has the right of pre-emption, and, failing it, the University of Buda-Pest.[1]

V. THE LAW OF FRANCE.

THE *CODE CIVIL*.

According to the *Code Civil*,[2] the property in a treasure found in a man's own land belongs to himself; if found in another person's land, one moiety belongs to the finder and the other moiety to the owner of the land. Treasure is defined as everything concealed or hidden which no one can establish to be his property, and which is found purely by the effect of chance. These terse and apparently explicit rules have, however, been the subject of much litigation. An article found on the surface, not being either concealed or hidden, is not treasure but waif (*épave*), and is regulated by other provisions of the Code, but courts and lawyers take different views as to the section that is applicable. The State has claimed as in right of masterless goods (*biens vacans*), but apparently without much success. The aid of the ancient Coutumes has been invoked, but in vain. These, it is held, are all superseded by the Code. The city of Paris and other bodies have attempted to make municipal regulations, but these have been challenged as *ultra vires*, being contrary to the fundamental provisions of the Code, which alone is law.

The partition between finder and owner does not apply in the case of excavations undertaken for the purpose of making discoveries, nor to valuables found in a tomb in a field. A treasure concealed in a wall is held as appurtenant to the land and not as an accessory to the building. If, therefore, the proprietor sells the materials and a treasure is found during its demolition, it belongs to the owner of the site, not to the purchaser of the material.

[1] Wussow, *Die Erhaltung der Denkmäler*, i., p. 209.
[2] *Code Civil*, § 716: cf. §§ 539, 713, 717, 768; Laurent, *Principes de droit civil Français*, T. viii., § 447 *et sqq.* (Paris, 1873).

THE OLD LAW.[1]

64. With regard to treasure hidden in the earth, in a field, or in a house, the question has been raised whether it ought to belong *jure inventionis* to the finder, or to the proprietor of the field or house where it was found, *jure accessionis*, as a pertinent. The Roman lawyers determined the question by giving one half of the treasure to the finder and the other half to the owner of the field.[2]

By the French law treasure is divided between the superior having baronial jurisdiction (*le seigneur haut-justicier*) in the territory in which the treasure was found, the owner of the place where it was found, and the finder, each taking one third.

65. The finder of a treasure has right to a share only if he has found it by chance; as, for instance, when a man, in digging a ditch in a field by order of the owner, finds a treasure, or when a cleanser of wells or sewers finds a treasure in them. But if without the consent of the proprietor one has made excavations in a field in search of treasure, and has been successful, the single law of the Code *de thesauris* ordains that in this case he shall not have a share, because he ought not to profit by his own wrong in digging in another's field without his consent.

66. Note that by treasure we understand a thing of which we have not any indication to whom it formerly belonged: *Thesaurus*, says Paul, *est vetus quaedam depositio pecuniae, cujus non extat memoria, ut jam dominum non habeat.*[3] But if there be any indication or presumption which points to the person who has concealed the money or other object in the place where it was found, it ought not in this case to be considered as treasure, but belongs to the person who hid it or to his heirs, to whom the finder must deliver it, *Alioquin*, adds Paul, *si quis aliquid vel lucri causa, vel metus, vel custodiae, condiderit sub terra, non est thesaurus, cujus etiam furtum fit.*

Scaevola gives this example: *A tutore pupilli domum mercatus, ad ejus refectionem fabrum induxit; is pecuniam invenit. Quaeritur,*

---

[1] Pothier, *Traité du droit de Propriété*, i. 2, 4, § 2.
[2] *Inst. tit. de rer. divis.* § 39 (*Inst.* 2, 1, 39).
[3] *L.* 31, § 1, ff. *De acq. rer. dom.* (*D.* 41, 1, 31, § 1); see *supra*, p. 61.

ad quem pertinent? Respondi: si non thesauri fuerunt, sed pecunia forte perdita, vel per errorem ab eo, ad quem pertinebat, non ablata; nihilominus ejus eam esse, cujus fuerat.[1]

Si non thesauri fuerunt, that is to say, if it did not appear that this money was treasure, as would be the case if it consisted of ancient coins, which appeared to have been put in the place where they were found at a remote time, and in such a manner as made it impossible to know by whom they had been deposited. If, on the other hand, it should appear that the deposit was recent, as, for instance, because it consisted of pieces of modern issue, the presumption in this case is that the money had been placed there by the father of the minor when he occupied the house; that this money was only mislaid; that it was by mistake that the tutor in selling the house had not removed it from the place for want of knowing where it was; and that this money, having always continued to belong to the father of the minor, must be restored to the tutor of the minor as heir.

VI. THE LAW OF SPAIN.[2]

THE CIVIL CODE.

348. Property is the right to enjoy and to dispose of a thing, without any restrictions other than those imposed by law.

The proprietor can sue the possessor and holder for the purpose of reclaiming it.

349. No one can be deprived of his property, except by competent authority and for a proved case of public advantage and on receiving full compensation.

If such compensation has not been paid the judges will maintain and, if need be, reinstate the owner in possession.

350. The proprietor of a piece of land is owner of what is on it and under it; he can make such works, plantations, and excava-

---

[1] *L.* 67 ff. *De re vind.* (*D.* 6, 1, 67).

[2] A Civil Code for Spain was promulgated upon 24th July, 1889. It does not, however, supersede the *fuero* or customary law, which is allowed to remain (*Code* § 12), so that in fact the new law, in a manner, merely supplements the old.

The Code is based to a very large extent on that of France, but the arrangement is more scientific and the draftsmanship better. It is a clear and methodical statement of the general principles of law.

tions as he thinks proper, having regard to existing servitudes and conforming to the laws regarding mines and water and police regulations.

351. Hidden treasure belongs to the owner of the land in which it is found.

If, however, the find be made in the property of another person or of the State, and by chance, the half goes to the finder.

If the objects found are of scientific or artistic interest, the State may acquire them at their just value, which will be apportioned as aforesaid.

352. By treasure is meant, in this connection, a secret and unknown hoard of money, of trinkets or other precious objects, the lawful proprietorship of which is not established.

614. Whosoever, by chance, finds a treasure in the property of another will have the right granted to him by article 351.

615. The finder of a moveable, which is not treasure, must hand it over to its former possessor. If he is unknown he must deliver it to the Alcalde of the place where it was found.

The Alcalde will give public notice of it in ordinary form on two consecutive Sundays.[1]

If the article cannot be kept without loss or without expense disproportionate to its value, it will be sold by auction on the expiry of eight days after the second notice if the owner has not appeared, and the proceeds will be deposited.

After the lapse of two years from the second notice, if the owner has not appeared, the article found, or the sum realised by its sale, will be handed to the finder. The latter, in the same manner as the owner, is bound to pay the expenses.

616. If the owner appears in time, he shall be bound to reward the finder with a tenth part of the value of the article found, or its price if sold. If the value exceeds 2000 *pesetas* the reward as respects the surplus will be reduced to a twentieth.

---

[1] This was the practice under the old French customary law. Thus by the Coutume d'Orleans, *art.* 65, "Strays (*épaves*) must be proclaimed on three several Sundays, at sermon at parish high mass, and in the place where they were found, on court days, at the instance of the barons—greater or less (*seigneurs de haute, moyenne et basse justice*)—or of the finder of the strays. Pothier, *Traité du droit de Propriété*, i. 2, § 3.

THE OLD LAW.[1]

"The treasure which is found upon the earth, or concealed in it by any one, is applied to the benefit of the Exchequer (*al fisco*), with a reservation of the fourth part for the finder, who ought to communicate the discovery to the justice.[2]

VII. THE LAW OF DENMARK.

THE CODE OF CHRISTIAN V.[3]

CHAPTER IX.—OF THINGS FOUND.

ARTICLE 3. Whatever gold or silver is found in the hills or in plowing or otherwise, is the property of the king alone, and is called the Treasure of Denmark.

VIII. THE LAW OF RUSSIA.[4]

Treasure, by which is meant objects of value hidden in the earth or in a wall, belongs, according to the law of Russia proper (*Svod Zakanov*), to the owner of the land, without distinguishing between the case where the owner has himself been the finder and that of a stranger having found it. Under this code no private person and no local authority can undertake or authorize excavations on the land of another without his consent.

---

[1] Johnston, *Institutes of the Civil Law of Spain*, p. 101 (London, 1825, 8vo).

[2] L. 3, tit. 22, Lib. 10, *Novísima Recopilacion de las Leyes de España*; L. 1, tit. 13, Lib. 6, *Leyes de Recopilacion*, which alters L. 45, tit. 28, P. 3, *Leyes de Recopilacion*. See Lagunez, *De Fructibus*, pars. i., cap. 11 (Lugduni, 1702, fol.).

[3] *The Danish Laws, or the Code of Christian the Fifth, faithfully translated for the use of the English inhabitants of the Danish Settlements in America*, p. 334 (London, 1756, 8vo).

[4] The general law of Russia is contained in the *Svod Zakanov* published in 1833, and in force since 1st January, 1835, subject to periodical amendment. The kingdom of Poland is still substantially subject to the Code Napoléon, which was introduced in 1808. In the Baltic Provinces their own customary law prevails. It was reduced to order under the direction of Alexander II. (*Liv-est-und curlaendisches Privatrecht*) and published in 1864 (St. Petersburg, 1864, 4to). The Grand Duchy of Finland, which was only separated from Sweden in 1809, is still governed by the Swedish Code of 1734 (*Sveriges Rikes Lag ed. Sjöros et Sulin*, Helsingfors, 1874, 8vo).

In the Governments of Tchernigof and of Poltava, in Poland, as well as in the Baltic Provinces, when a treasure has been found by chance on the property of another, a private person, a corporation, or the Crown, it is divided between the owner of the land and the finder. If it be found by the landowner himself it belongs wholly to him. The case is the same where the find is due to excavations undertaken by a third person, either by authority of the landowner or without his knowledge. On the other hand, when treasure is found on land which has no owner, the finder alone is entitled to it (*Code of the Baltic Provinces*, 745 *seqq.*). The legislator for the Baltic Provinces had a difficulty in deciding, in the case where the *dominium utile* of the land had been separated from the *dominium directum*, whether treasure trove should belong to the vassal or to the superior, but following local customs he decided in favour of the vassal (*Ib.* 950), a decision which is considered not free from criticism.

The same code determines that objects which have been hidden or buried, but which are not technically treasure, either because it is impossible to say to whom they should go, or because they have been found in a chattel, are to be assimilated, in so far as concerns appropriation, not to treasure trove, but to strays or finds of ordinary articles. The owner of an article which has been found is not bound to recompense the finder if he proves that he knew where it was concealed[1] (*Ib.* 742, 743, 748).

IX. THE LAW OF TURKEY.[2]

According to Mussulman law, if discovered treasure bears an Islamite mark, it is ranked amongst objects lost and found (*loqta*), and in this case it is subjected only to the relative dues; but if it bears an emblem of infidelity, such as the figure of an idol or of a cross, it will be subjected to "Khoums." According to the regulations applicable to things lost and found (*loqta*) no inanimate object found, worth more than one dirhem, can be the object of a use or of a location without the consent of its legitimate owner.

---

[1] Lehr, *Eléments de droit civil Russe*, p. 240 (Paris, 1877, 8vo).

[2] Ongley, *The Ottoman Land Code*, pp. 58, 335 (London, 1892, 8vo).

LAW AS TO TREASURE TROVE

Sometimes, however, "whatever is found in the desert, or buried in the earth, or in the intestines of animals, or in the bosom of waters, without its being possible to know the owner, becomes the entire property of the finder."[1] "Khoums," to which the non-Islamite treasure is subjected, literally means "the fifth part," which, in the cases laid down by the law, the Mussulmans must allow to be deducted as tax from their patrimony.[2]

[1] Tornauev, *Droit Mussulman, Traduit en français par Eschbach*, p. 283 (Paris, 1860, 8vo).

[2] *Op. laud.*, 61, 62. There is a special law relating to excavations and search for coins, etc. (22 Rebr-ul-Akher 1301, *i.e.* $\frac{9}{21}$ Feb. 1884), Wussow, *Die Erhaltung der Denkmäler*, ii., p. 314.

www.ingramcontent.com/pod-product-compliance
Lightning Source LLC
Chambersburg PA
CBHW020137170426
43199CB00010B/785